T0224609

Lecture Notes in Computer Science 1748

Edited by G. Goos, J. Hartmanis and J. van Leeuwen

Springer

Berlin
Heidelberg
New York
Barcelona
Hong Kong
London
Milan
Paris
Singapore
Tokyo

Hong Va Leong   Wang-Chien Lee
Bo Li   Li Yin   (Eds.)

Mobile
Data Access

First International Conference, MDA'99
Hong Kong, China, December 16-17, 1999
Proceedings

Springer

Series Editors

Gerhard Goos, Karlsruhe University, Germany
Juris Hartmanis, Cornell University, NY, USA
Jan van Leeuwen, Utrecht University, The Netherlands

Volume Editors

Hong Va Leong
Department of Computing, The Hong Kong Polytechnic University
Hung Hom, Kowloon, Hong Kong, China
E-mail: cshleong@comp.polyu.edu.hk

Wang-Chien Lee
GTE Laboratories Incorporated
40 Sylvan Road, Waltham, MA 02451, USA
E-mail: wlee@gte.com

Bo Li
Li Yin
Department of Computer Science
The Hong Kong University of Science and Technology
Clear Water Bay, Kowloon, Hong Kong, China
E-mail: {bli;yinli}@cs.ust.hk

Cataloging-in-Publication data applied for

Die Deutsche Bibliothek - CIP-Einheitsaufnahme

Mobile data access : first international conference ; proceedings / MDA '99,
Hong Kong, China, December 16 - 17, 1999. Hong Va Leong ... (ed.). - Berlin ;
Heidelberg ; New York ; Barcelona ; Hong Kong ; London ; Milan ; Paris ;
Singapore ; Tokyo : Springer, 1999
  (Lecture notes in computer science ; Vol. 1748)
  ISBN 3-540-66878-0

CR Subject Classification (1998): C.2, C.5.3, C.3, D.2, H.5, H.4

ISSN 0302-9743
ISBN 3-540-66878-0 Springer-Verlag Berlin Heidelberg New York

© Springer-Verlag Berlin Heidelberg 1999
Printed in Germany

Typesetting: Camera-ready by author
SPIN: 10750047   06/3142 – 5 4 3 2 1 0   Printed on acid-free paper

Preface

With the rapid development in wireless-network and portable computing and communication devices, mobile users are expected to have access to information from anywhere at anytime in the near future, in the form of ubiquitous computing, a term coined by the late Mark Weiser of Xerox, PARC. Indeed, the emerging mobile technology will probably bring us the next wave of information revolution and change our society as we move into the next millennium. Before this vision can be realized, a number of challenges have to be overcome. Traditionally, network-based information systems have been developed under wired assumptions about the connectivity and topology of the underlying networks. To eliminate these limitations from wireless and mobile environments, research efforts are needed in networks, architecture, software infrastructure, and application levels, in order to provide mobile data access over hybrid wireless and wired networks, which is the central theme of this conference.

These proceedings collect the technical papers selected for presentation at the First International Conference on Mobile Data Access (MDA'99), held in Hong Kong, following its return to China, on December 16–17, 1999. The conference is held in conjunction with the International Computer Science Conference, the International Conference on Real-time Computing Systems and Applications, and the Pacific Rim International Symposium on Dependable Computing, forming part of the International Computer Congress.

In response to the call for papers, the program committee received 39 submissions from North America, Europe, Asia, and Oceania. Each submitted paper underwent a rigorous review by three to four referees, with detailed referee reports. Finally, these proceedings represent a collection of 20 research papers, with contributions from ten different countries from four continents. The contributed papers address a broad spectrum on mobile data access issues, ranging from the lower level core research efforts on communication on wireless networks and location management, to the intermediate level research topics on data replication and transaction processing, and finally to the higher user level research applications on ubiquitous information services.

We are extremely excited to align a very strong program committee, with outstanding researchers in the areas of mobile computing and databases. We would like to extend our sincere gratitude to the program committee members, who performed a superb job in reviewing the submitted papers for contributions and technical merits towards mobile data access.

Last but not least, we would like to thank the sponsors, for their support of this conference, making it a success. Thanks go to the IEEE Hong Kong Computer Section, ACM Hong Kong Chapter, Sino Software Research Institute at Hong Kong University of Science and Technology, the IEEE Technical Committee on Personal Communications, and ACM SIGMOBILE.

December 1999   *Hong Va Leong, Wang-Chien Lee, Bo Li and Li Yin*

The First International Conference on Mobile Data Access (MDA'99) In Conjunction with the International Computer Congress 99 (ICC'99)

Executive Committee

General Chair:	Dik Lun Lee (Hong Kong University of Science and Technology)
Organizing Chair:	Kam-Yiu Lam (City University of Hong Kong)
Secretary:	Dik Lun Lee (Hong Kong University of Science and Technology)
Treasurer:	Man-Hoi Choy (Hong Kong University of Science and Technology)
Local Arrangements:	Edward Chan (City University of Hong Kong)
	Victor Lee (City University of Hong Kong)
Publication:	Hong-Va Leong (Hong Kong Polytechnic University)
Technical Program Co-Chairs:	Wang-Chien Lee (GTE Labs, USA)
	Bo Li (Hong Kong University of Science and Technology)

Technical Program Committee

Swarup Acharya	Bell Labs, Lucent Technologies (USA)
Badri. R. Badrinath	Rutgers University (USA)
Daniel Barbara	George Mason University (USA)
Arbee L.P. Chen	National Tsing Hua University (Taiwan)
Ming-Syan Chen	National Taiwan University (Taiwan)
Quan Long Ding	Centre for Wireless Communications (Singapore)
Pamela Drew	Boeing (USA)
Michael Franklin	University of Maryland (USA)
Lixin Gao	Smith College (USA)
Sandeep Gupta	Colorado State University (USA)
Zygmunt J. Haas	Cornell University (USA)
Sumi Helal	University of Florida (USA)
Qinglong Hu	University of Waterloo (Canada)
Nen-Fu Huang	National Tsing Hua University (Taiwan)
Shengming Jiang	Centre for Wireless Communications (Singapore)
Dae Young Kim	Chungnam National University (Korea)
Kin K. Leung	AT&T Labs - Research (USA)
Ping Li	City University of Hong Kong (Hong Kong)

Yi-Bing Lin	National Chiao Tung University (Taiwan)
Xiaomao Liu	Florida International University (USA)
Michael R. Lyu	Chinese University of Hong Kong (Hong Kong)
Radhakrishna Pillai	Kent Ridge Digital Labs (Singapore)
Evaggelia Pitoura	University of Ioannina (Greece)
Krithi Ramamrithan	University of Massachusetts (USA) / Indian Institute of Technology, Bombay (India)
Mukesh Singhal	NSF/The Ohio State University (USA)
Krishna Sivalingam	Washington State University (USA)
Mallik Tatipamula	Cisco Systems (USA)
Eric Wong	City University of Hong Kong (Hong Kong)
Naoaki Yamanaka	NTT (Japan)
Lawrence K. Yeung	City University of Hong Kong (Hong Kong)
Stanley Zdonik	Brown University (USA)
Zhensheng Zhang	Bell Labs, Lucent Technologies (USA)
Moshe Zukerman	University of Melbourne (Australia)

External Reviewers

David E. Bakken	Chia-Min Lee
Jihad Boulos	Anthony Lo
Jyh-Cheng Chen	Sanjay Madria
Yonghong Chen	Masayoshi Nabeshima
Amir Djalalian	Eiji Oki
Lyman Do	Christine Price
Jun Du	Dan Rubenstein
Chuan-Heng Foh	Subhabrata Sen
Dajiang He	Naoki Takaya
Stuart Jacobs	Ning Yang
Jin Jing	

Sponsoring Institutions

Sponsored by IEEE Hong Kong Computer Section, ACM Hong Kong Chapter, and Sino Software Research Institute (Hong Kong University of Science and Technology)
Technical Sponsorship from IEEE Technical Committee on Personal Communications and in cooperation with ACM SIGMOBILE

Table of Contents

Session IV: Mobile Data Replication and Caching

Session V: Mobility and Location Management

Keynote Speeches

Session I: Wireless Networks and Communications

Wireless VoIP: Opportunities and Challenges

Jin Yang and Ioannis Kriaras

UMTS Advanced Development, Lucent Technologies
Windmill Hill Business Park, Swindon SN5 6PP, United Kingdom (jinyang,
ikriaras@lucent.com)

Abstract. Next generation wireless networks are going to use IP capable mobile terminals and support both voice/multimedia and data services. VoIP is a promising technology allowing a converged core network
for all services, and provides a service platform for easier and quicker
service creation. Application of VoIP in wireless networks, however, is
not so straightforward as people may expect. Among other issues, mobility support is perhaps the most challenging one. The different levels
of mobility support (terminal, personal and service mobility) highlight
the gap between the existing VoIP framework, which is so far focusing
on fixed networks, and the requirements in a wireless environment. After
analyzing the existing solutions/proposals for this problem, a framework
is proposed in which the mobility-related services are decomposed into
several levels so that mobility support is made available to many applications. This layered approach would help to lead to a mobility enabled
wireless IP infrastructure in which voice service is just an application.

1 Introduction

IP (Internet Protocol) is going to be the common platform for both data and
voice/multimedia services in the near future. VoIP (voice over IP) starts from
making and receiving voice calls using Internet. The scope of VoIP is then extended to multimedia services and interworking with PSTN (public switched
telephone network) including IN services. However, most of the work in this
area so far is done with a wireline network as the default targeting architecture and much less attention has been paid to the app lication of VoIP in a
wireless/mobile environment. This situation just starts to change now - VoIP
in wireless networks becomes a topic attracting attentions in both VoIP world
and wireless industries, particularly in the light of 3G (third generation) wireless
networks. The intention of this paper is to investigate the issues and challenges
of applying VoIP in wireless environment, with focus on the different levels of
mobility support.

The reminder of this paper is divided into three parts. In section 2 the concept of wireless VoIP is introduced and some issues and problem of applying
VoIP technologies in wireless networks are briefly discussed. Section 3 analyses
the requirements and existing solutions for mobility support in VoIP. Section 4
presents a layered approach for mobility support in wireless VoIP. The paper is
concluded in Section 5.

H.V. Leong et al. (Eds.), MDA'99, LNCS 1748, pp. 3–13, 1999.
© Springer-Verlag Berlin Heidelberg 1999

2 Wireless VoIP

Wireless systems, e.g. GSM, are so successful that the number of subscribers is foreseen to match the number of wireline subscribers. To further extend the successful story of 2G (second generation) system, next generation (3G) wireless systems are under development, which would provide not only telephony but also multimedia and other data services over IP. The customers are certainly expecting not only all the benefits of 2G to be available, but also more new and/or cheaper services and features. The competition lies mainly in an efficient network infrastructure and rich service features.

On the other hand, VoIP allows a common packet backbone network for both voice and data services and is in favor of quick service creation. Some of the benefits of using VoIP include, from operators' viewpoint:

- Efficient use of network resources due to the statistical multiplexing between voice sessions and/or between voice and other applications.
- One single core network to manage and maintain.
- Lower cost of transport and service infrastructure.

From user and service provider's viewpoint, using VoIP provides:

- Multimedia support. The service quality can also be different, e.g. phone call of either CD quality or low-bitrate can be supported.
- Easier/quicker introduction of new services. For example by combining voice with other applications, e.g. web based call center. Multiple (third party) service providers can be available and there is no need to wait until the operator of today to introduce them. Also a better GUI is available.

There is therefore a very good opportunity for these two rapidly developing technologies, 3G wireless and VoIP, to work together. There are various architectural options for deploying VoIP technologies in wireless networks. For example:

a) End-user VoIP. In this case VoIP is running at the user's wireless terminal as an IP application. The underlying wireless network can be packet based, e.g. wireless LAN, GPRS systems, or circuit based, i.e. IP over circuit data.
b) Network VoIP, in which IP (or packet switch) network is used inside the wireless system for transport of user traffic and control signal. For example, IP interface at base station, IP based core network like in GPRS and UMTS. The use of IP and VoIP is transparent to end-users.

Wireless VoIP, in this paper, refers to the first type scenario, as shown in Fig. 1, in which the mobile terminal is IP capable and acts as a VoIP end point, e.g. a H323 terminal. 3G wireless systems, e.g. IMT-2000 families, UMTS and CDMA-2000, are targeting networks for this study.

A straightforward solution for wireless VoIP can be: wireless operators provide "IP pipe" with IP mobility support, for example based on mobile IP, and apply VoIP technology over it. In this solution the fact of wireless user(s) being involved becomes transparent. However such combination implies a series of challenges. Some of them do not exist in fixed networks or are not as obvious or critical as in the wireless environment, for example:

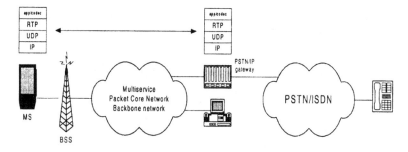

Fig. 1. Wireless VoIP: applying VoIP at wireless terminals

- **Radio resource efficiency**. Bandwidth is typically more expensive in cellular radio networks than in the fixed ones. When voice/media packets are transported over RTP/UDP/IP layers, the combined headers of these IP layers dramatically reduce the radio efficiency. For example, when GSM FR codec is used, the combine headers, at least 40 bytes, would be nearly the double of the payload (23 bytes). Header compression in wireless network presents extra complexity [10].
- **QoS support**. QoS is always a serious concern. The issues include: the impact of mobility on QoS management; Position of vocoder; How to map QoS requirement down to lower layer QoS support in wireless network; and how to integrate with QoS mechanism in IP backbone network.
- **Heterogeneous networks**. 3G systems allow speech services to be provided in circuit mode or packet mode. The network can be all-IP based or a hybrid of circuit-switch and packet-switch system. 3G terminals may support only circuit or packet capability or both. Supporting roaming and handover between these networks using different type of terminals is a serious challenge.
- **Mobility support**. Mobility is not only an issue for wireless users. However mobility support is one of the most important features for wireless users. There are different levels of requirements for mobility support, which are further discussed in the following sections.

The above issues, among others, have to be addressed before wireless VoIP can be a competitive and appealing solution and be successfully deployed in next generation wireless systems.

3   Mobility in VoIP: Requirements and Existing Solutions

There are different levels of mobility support [8, 9]:
- **Terminal mobility** allows moving a terminal, i.e. a telephone, laptop, PDA etc., from one location to another while maintain active communication.

- **Personal mobility** is the ability of end users to originate and receive calls on any terminal in any location, and the ability of the network to identify the users as they move. Personal mobility is based on the use of a unique personal identity.
- **Service mobility** is the ability of end users to access subscribed telecommunication services across network, which have the same look and feel independent of the current network and terminal in use.

These requirements are not new. Actually many of them have been realized in existing wireless systems e.g. GSM, and a full solution is under discussion in telcom world. These requirements need to be fulfilled in IP and VoIP platform as well.

Recently mobility support in VoIP has attracted many attentions, particularly from wireless industry. In this section we would like to review and summary this work.

Terminal Mobility
The terminal mobility can be based on IP layer mobility or as an integrated part of VoIP applications.

Mobile IP
The well-known Mobile IP standard [6] provides a solution for global roaming capabilities to IP hosts, through the use of "Home Agent" and "Foreign Agent". It also allows maintaining the on-going application-level session in a transparent way. There are some serious issues and limitations when applying Mobile IP into global wireless environment with high mobility [4]. One of them is the signaling overhead, as Mobile IP does not define paging areas so the frequent location update messages need to travel across "core" network. Another issue is the handover latency, caused by the long distance signaling, which is critical for real-time applications like VoIP. Some proposal have been made around a hierarchical approach that separates global mobility, provided by Mobile IP, from "local" mobility that handles micro-mobility and avoids the need for long-distance interactions with home network. The micro-mobility mechanism can be based on, for example, "cellular IP" [7], o r access system specific mobility support, e.g. GPRS and wireless LAN layer 2.

Mobile Extension of H323
ITU H.323 [5] is the widely deployed standard for VoIP so far. H.323 specifies technical requirements for multimedia communications, including the system components, control messages and functions for component communications, over packet switched networks. An approach has been proposed [3] to extend H.323 that combines the characteristics of both cellular phone system and mobile IP mechanism with Internet telephony, and therefore realizes the transmission of real-time voice traffic for both stationary and mobile hosts over IP-based networks. The feature of the proposed approach is that, when the terminal moves,

the complicated mobility management functions can be handled by the procedures of dynamically joining and departing from a conference, which are functions already defined in H.323. Therefore, the approach allows mobility support without the need for additional new entities and with minimal modifications to existing H.323 standard.

Personal Mobility

For personal mobility support, VoIP call control system needs to be aware of the user's location, e.g. network address, and routes calls correspondingly. One way of doing this is by capturing the user location information during user registration process. For example, in SIP (session initiation protocol [11]), users register to a registrar server by sending REGISTER requests, which allows a client to let a proxy or redirect server know at which address he/she can be reached so that user roaming ca n be supported. In [1, 2] a general model is suggested for personal mobility support in a H.323 framework, allowing use of different type of terminal.

Service Mobility

One possible approach is to address all call signaling to the home network/service provider, and execute the service logic and call control there (media data does not necessarily goes through home network). The visiting network provides IP connectivity only and is unaware of the VoIP calls. There are some issues for this approach. For example, some services, e.g. emergence call, need to be handled locally. Another issue is that in many cases local VoIP facilities, e.g. PSTN/IP gateway, are better used to ac hieve low-cost and better QoS, which requires the co-ordination between the home and visiting network domains.

3.1 Experiences from Cellular System Design

The design and success of cellular systems can provide us with valuable hints and help in solving the mobility issues in VoIP framework. The major components and procedures of the mobility management in GSM system are briefly reviewed in the following.

Location Management. The location of a GSM mobile is managed using two important entities: HLR (Home Location Register) and VLR (Visitor Location Register). When a mobile station is switched on in a new location area, or it moves to a new location area or different operator's network, it must register with the current network to indicate its current location. At the same time HLR sends subscriber information, needed for call and supplementary service control, to the VLR.

Call Routing. An incoming mobile terminating call is directed to the GMSC (Gateway Mobile Switching Center) function. The GMSC is basically a switch, which is able to interrogate the subscriber's HLR to obtain routing information. Normally, the HLR knows which network should receive incoming calls for the roaming mobile and the VLR knows the exact location of the mobile and which BTS should page the mobile.

Handover. GSM is designed with sophisticated procedures to support different levels of handover. The network components, with the help from mobile terminal, monitor and measure the signaling strength when a terminal is moving. When a handover is necessary, the resources in the new cell are allocated and the mobile is informed to switch to new cell. The terminal mobility is realized at the link layer.

SIM card. The mobile station consists of a terminal and a subscriber identity module (SIM) that provides a certain level of personal mobility so that the user can have access to subscribed services irrespective of a specific terminal within the GSM systems. By inserting the SIM card into another GSM terminal, the user is able to receive calls at that terminal, make calls from that terminal, and receive other subscribed services. The usage of SIM card is restricted to the GSM terminals.

Service mobility is provided to the user through Intelligent Network (IN) platform (e.g. using CAMEL) within the GSM systems where the end user should not see any difference in the services provided by the IN nodes irrespective of the user's location and terminal used.

4   Mobility Support for Wireless VoIP: a Layered Approach

4.1   Vertically Integrated Model vs. Layered Approach

GSM, for example, provides a complete solution for mobility support, as summarized in 3.1. As other conventional telecommunication systems, GSM is voice centric: voice services are seen as one of the "basic services" of the system, which is elemental and indivisible. The voice services, including supplementary services in GSM, are vertically integrated with other parts of the network and are standardized. The different levels of mobility support are tightly integrated in a monolithic way. As a result, GSM system has achieved highly optimized and reliable support for voice services. Mobility support is built deeply into the system architecture.

In the IP world, however, voice services can be seen and supported as just another application on the IP platform. There is a clear separation at protocol and structural level between IP connectivity and applications. This vision implies that the network should not be designed to support a single solution for voice service. Instead the IP platform should provide infrastructure so that various applications and services, including voice service, can be easily and quickly created. This is the approach we would like to follow in working out the solution for wireless VoIP mobility support.

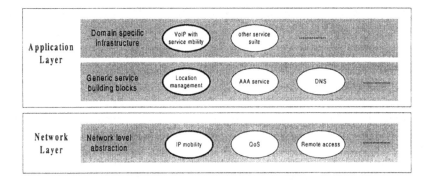

4.2   An Environment in which Wireless VoIP Can be Supported

One might attempt to extend H.323 suite to cope with all levels of mobility requirements. However the following questions need to be answered:
- Is this solution working only with certain type of wireless IP networks.
- Is this solution for VoIP only?
- Is this solution valid only for a particular call control protocol, e.g. H.323 ?
- Can this solution help other applications in mobility support?

Instead of developing a single integrated solution, we would prefer to a layered approach aiming for a service environment in which mobility related services are provided at different levels. Thus VoIP is just an application in this environment.

The proposed mobility enabled IP environment consists of several levels of mobility related services, as illustrated in Fig. 2: IP layer mobility, user location service and application mobility. The position of each service is determined by the generality of the provided service. A high level framework is discussed below. The detailed design of each service layer including the service access interface and protocols are left for further study.

IP Layer Mobility
VoIP represents a class of applications on IP platform and obviously IP layer mobility is a key feature. IP layer mobility provides network layer solution which allows an IP host to move and may change its IP access point but keeps upper layers, e.g. transport and application layers, untouched. The focus of this level service is mobile IP hosts. There is no single best solution for IP mobility, which is usually determined by various factors e.g. the size of the network, the access system etc.. The possible solutions, for example, are:
- Mobile IP, as defined in [6].
- Hierarchical IP mobility management such as cellular IP [7].
- Wireless access system specific solution for IP layer mobility support. For example GPRS (General Packet Radio System) defined its own mobility mechanism for IP hosts.

- Use of lower layer mobility solution. For example, dial-up access over GSM link, Wireless LAN with layer 2 mobility support.

IP mobility is a well-known issue and there is a lot of work on going particularly from wireless perspective [7]. It is a general requirement for all IP applications in a mobile environment.

User Location Management Service

Similar to DNS (domain name system), which maps between IP host names and their IP addresses, user location management service captures the user location information and maps a personal identifier to a network address at which the user can be contacted. The personal ID can be in the form of E164 address (telephone number), email address, or other formats. The location information can be represented by the access network type and network addresses. For example, the location of a wireless IP user is the IP a ddress of the terminal being used. When roaming to GSM network, his location is the telephone number (MSISDN) of the terminal (a dual-mode terminal or a separate one).

As far as IP user is concerned, there are two granularity of the location service: one is to locate the user using his home IP address, which is more static and easier to provide. In this case, the location service maps between user ID and home IP address and relies on IP layer mobility when the user is roaming. Thus some inefficiency, for example caused by mobile IP, will be unavoidable.

Another one is to associate a user with his IP address in the visiting network, which reflects a more accurate location mapping. Furthermore, for wireless users, their geographical locations can be part of user location information and can be used for many location-based services. There is no standard way for capturing the location information. It is more likely that various sources are used for this purpose by exchanging location information among several network entities. For example, some access systems may provide build-in location tracking mechanism, e.g. HLR, or HA (home agent) in mobile IP, which can be exposed to different applications via the location service API. Also it is possible to share the location information captured by other applications, e.g. by VoIP applications during user registration.

Updating location information is another issue. In many cases, the network address does not change frequently along user's movement. For example in GPRS, the dynamically assigned IP address does not change during a PDP session. When it changes, the location server needs to be updated.

It is unlikely that the location management is a centralized entity. Instead it will be distributed and run by different domains. A protocol needs to be provided so that different location server can exchange location information.

Application Level Mobility

This level of mobility addresses the requirement of service mobility in a particular application domain. For example, as far as VoIP is concerned the service mobility means that a user can access the subscribed services regardless the location and the access network being used. VoIP itself is a service platform on which new voice/multimedia related services can be developed.

For VoIP applications, one solution for service mobility, as discussed in Section 3, is to ask the home VoIP provider to handle all call control, including call features. This approach has the benefit of simplicity and easy implementation. However, as the local visiting network is kept out of the service process, the local VoIP resources, e.g. the PSTN/IP gateway, may not be usable. Some services, e.g. emergency call, need to be handled by visiting network.

Borrowing experiences from GSM, we propose a possible solution for service mobility in VoIP. The major idea is:

- Allowing both home and visiting networks/VoIP service providers to participate call processing.
- Standardization of a small set of basic call services. Most calls are basic calls or requests the most-used three or four call features (e.g. call forwarding).
- Use of service profile. The service profile records user's subscription status and some service related data. When a VoIP user roams to a foreign network and prefers the local provider to handle his calls, his service profile can be transferred.
- Separation of call control and service control. Visiting network is responsible for basic call control and handling of standardized call features. For service provider specific calls the home network is contacted for service control.

4.3   An Example Scenario

In Fig. 3, an example scenario is presented showing how the three levels of services are used in order to make a VoIP call to a user who is using a mobile terminal.

The major procedures are:

1) The MS registers with the CC/SC (call control/service control) server in the visiting network, including its user ID and current IP address.
2) The user location information is captured by ULM (user location management) in the visiting network.
3) The location information is propagated to the ULM in the user's home network, which can be used for other applications as well, for example news updating service.
4) A call coming from PSTN to the roaming user.
5) The caller's network captures the call request and sends the request to the callee's home domain CC/SC server.
6) The home CC/SC server checks the user's service profile and, if the user is entitled for the service, contacts ULM for user location information.

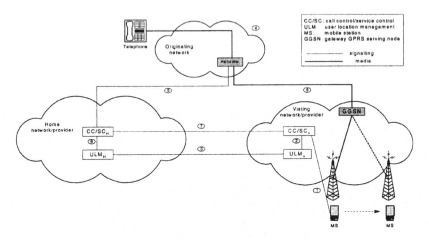

Fig. 3. An example application scenario

7) The home CC/SC sends the call request to the visiting network CC/SC, which tries to establish a VoIP call between the terminal and a PSTN user via a PSTN gateway. To achieve service mobility it may contact home SC during call initiation and termination stages for service control. The details of these procedures are not shown.

8) After call setup, media traffic is transported over IP between the PSTN gateway and MS via GGSN (a GPRS/UMTS entity). The MS can move while keeping active communication. The terminal mobility is provided by the IP mobility in the access system.

5   Conclusion

In this paper we pointed out that the application of VoIP technology with wireless IP systems would allow more efficient network architecture and a better service creation environment. Such a combination implies a series of challenges. Among other issues, mobility support is a critical requirement for wireless VoIP to be successful. Instead looking for "one-solution-for-one-application", we proposed to decompose the mobility-related services into several layers, which can be the components of mobility ena bled IP platform.

This layered approach separates concerns and allows "divide-and-conquer" and encapsulation of implementation details. This approach also leads to a mobility enabled IP platform. Wireless IP platform is for every services and applications. VoIP is just one (class) of them. By decomposing the required services into different levels, many of them can be generalized and be useful for a wide-range of applications.

References

1. Jin Yang, Emiliano Mastromartino, "A proposed model for TIPHON mobility support", 10TD103, contribution to ETSI TIPHON, June 1998, Israel.
2. Jin Yang, Emiliano Mastromartino, Ioannis Kriaras, "Personal mobility support in VoIP", Proceedings of 4th ACTS Mobile Summit, June 1999, Italy
3. Wanjiun Liao, Victor O. K. Li, "Mobile extension to H323", 13TD092, contribution to ETSI TIPHON, May 1999, Thailand.
4. Guilhem Ensuque, "Towardts an integrated IP/UMTS network architecture", Proceedings of 4th ACTS Mobile Summit, June 1999, Italy
5. G. A. Thom, "H.323: the Multimedia Comm. Standard for Local Area Networks," IEEE Comm. Magazine, pp. 52-56, Dec. 1996
6. C. Perkins ed., "IP mobility support", IETF RFC 2003, Oct. 1996
7. Andras G. Valko, "Cellular IP: a new approach to Internet host mobility", Computer Communication Review, ACM press, Vol 29, Number 1, January 1999
8. Henning Schulzrinne, "Personal mobility for multimedia services in the Internet ", European Workshop on Interactive Distributed Multimedia Systems and Services, (Berlin, Germany), Mar. 1996.
9. "Analysis of existing roaming techniques applicable to TIPHON mobility services", DTR7001, ETSI TIPHON document.
10. Mikael Degermark et al. "CRTP over cellular radio links", Internet-draft, IETF, June 1999.
11. Mark Handley et al., "SIP: session initiation protocol", Internet draft, IETF, Sept. 1998

GSM Mobile Phone Based Communication of Multimedia Information: A Case Study

Y.S. Moon[1], H.C. Ho[1], and Kenneth Wong[2]

[1] Dept. of Computer Science & Engineering
The Chinese University of Hong Kong, Shatin, Hong Kong
[2] Dept. of Engineering, Cambridge University England, U.K.

Abstract. Through a practical application, this paper studies the capacity of the GSM network, a wireless mobile phone network widely used in Europe and some Asian countries, for transmitting multimedia data. The application is a client-server based automobile security system which monitors the interior and exterior environments of the moving automobile by a computer equipped a digital camera, a voice capturing microphone and a GPS (Global Positioning Systems) receiver. The captured data is sent to the server computer through the GSM network. The design and construction of the system are described with parameters of the equipment and the procedures of the experiments clearly shown. Results of the experiments show that a basic system can actually be implemented even given the narrow bandwidth of the GSM data transmission system.

1 Introduction

Distant delivery is always insecure in the sense that what is going on during the delivery can often hardly be known [3]. Very often, the delivery is delayed or the delivery does not even arrive at the destination at all. Such uncertainty during delivery must be minimized. This is particularly true for the delivery of important and valuable goods such as transfer of cash between banks. One simple solution is to keep track of the remote object. Traditionally, tracing a remote mobile object usually requires another object to follow the object being traced and report to the central control unit through certain communication media such as radio. Although this is in fact a waste of resources, it seems to be the only way of doing so. With the development of GPS, this situation changes. Tracing a remote object now becomes much easier than ever before, and much more accurate too. By attaching a GPS receiver to the object being traced, the exact position of the object can be retrieved and transmitted back to the central control unit and be known to the monitoring party.

Nevertheless, being able to know the exact location of the object being traced will sometimes still not be sufficient for determining what goes wrong when the object behaves abnormally. In worst case, we will not be able to detect anything that goes wrong when the object being traced moves along a path as predicted. As a result, it will be highly desirable if more information of the object

H.V. Leong et al. (Eds.), MDA'99, LNCS 1748, pp. 14–23, 1999.

being traced can be monitored together with the positional data. This project is therefore designed and developed in response to this question by providing additional video and audio data capturing capabilities. The data captured is transmitted from the remote object to the central station through the GSM mobile phone network. The GSM network is chosen as its signals cover more than 90

2   Systems

Basically, the whole system can be divided into two parts, one being the client while the other being a server. These two geographically separated entities communicate with each other so as to achieve the objective of tracing a remote object. The client system plays the major role of gathering information about the remote object, and the server system is mainly responsible for presenting the information gathered by the client in a neat and meaningful way. The major functionality of the client system and server system will be described in details in the following paragraphs respectively.

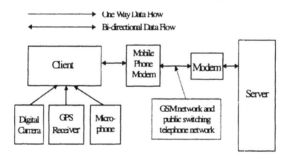

Fig. 1. Client/Server Model

2.1   Client System

The client system is the one to be attached to the remote object and is responsible for capturing all useful information about the remote object. This information includes the exact location of the remote object from the GPS receiver, as well as other data such as video and audio signals from other multimedia capturing devices. These captured signals will then be processed and integrated by the client system, which then sends these processed signals to the server system.

Since this client system is to be attached to the remote mobile object, it must be small in size and light in weight so as to increase its portability. Besides, since it is responsible for capturing several kind of signals and performing compression and integration on the data acquired, the client system must also have high speed and great processing power. Otherwise, smooth capture of data may not be possible and this may result in discontinuous data and loss of information, which in turns lead to poor accuracy and difficulties in comprehension.

A Pentium-233 MMX laptop computer running Windows 95 is employed here. It is equipped with a printer port for connecting a Connectix QuickCam black and white camera, a serial communication port for connecting a Garmin 45 GPS receiver and a PCMCIA port for connecting a mobile phone modem, the Ericsson Mobile Office DC 23v4. An Ericsson GH688 GSM mobile phone connects the modem to the GSM network. The modem confirms to ITU-T V.22bis and V.32 standards which facilitate data transfer at between 2,400 and 9600 bps. Besides, a signal booster is employed to amplify the signals of the mobile phone modem. Finally, a sound card connected with a microphone is also connected to the laptop for providing the sound capturing capability.

2.2   Server System

The server system is the one that will be located in a stationary place. The main task of the server system is to receive the captured signals from the client, process the received data and produce data visualization. Such data visualization includes displaying the path covered by the remote object on an electronic map and the corresponding latitude and longitude. The video and audio data received will also be processed to reproduce live video and audio signals.

The server also works as the command center and it has complete controls over the client system. It is responsible for sending command to the client to initiate or terminate its data capturing processes. The server system is also responsible for handling error during data transmissions.

A Pentium-133 MMX desktop computer running Windows 95 is employed here. It is equipped with a 14.4k bps modem for communicating with the client and a sound card with speakers for reproducing the audio data captured.

2.3   Cost

The following table shows the costs of some special equipment used in this project.

One point worth noting is that the prices of these equipment are dropping and the cost of building such a system will be expected to be much less in the future. Besides, the equipment could be compacted in to a box when the system is actually built for practical use.

3   GPS Signals

3.1   Interpolation

The GPS receiver needs at least signals from four different satellites to calculate its position accurately [3]. Very often, signals from the satellites will be blocked by high buildings. In such case, a GPS receiver may not be able to receive signals from four different satellites, and thus failing to calculate its exact position.

Table 1. Costs of some special equipment used

Item	Cost
Ericsson GH688 mobile phone	~US$350
Ericsson Mobile Office DC 23v4	~US$620
U.S. Robotics K56 voice modem	~US$150
Connectix B&W QuickCam	~US$130
Garmin 45 GPS receiver	~US$260
Signal Booster	~US$320
Total	~US$1830

Besides, since the data captured by the client system is transmitted to the server system via mobile phone networks, there are the possibilities of noise in the transmitted signals, or even worse, complete loss of signals when the object being traced enters into areas not covered by the mobile phone networks. In all the above situations, the server will not be able to obtain the location of the remote object. Instead of just displaying discontinuous path in the form of disjointed line segments on the map, the server will try to estimate the missing path of the remote object using the data obtained before and after the period of no data, together with the check points of a predefined path. The check points are some operator selected points along a path on the electronic map that the object being traced will follow. By joining these check points in order, the expected path of the object can be reconstructed.

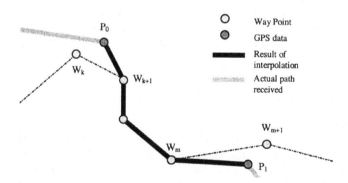

Fig. 2. Interpolation of GPS data

During a period when no GPS data is available, the server will then use the latest positional information P_0 before the loss period to find a line segment formed by two consecutive check points Wk and Wk+1 which is closest to P_0.

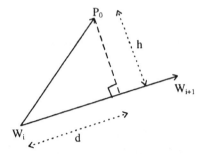

Fig. 3. Calculations involved in Interpolation

When the server receives GPS data from the client again, the server will then use the newly received positional information P_1, and the locations of two consecutive check points W_m and W_{m+1} closest to P_1 to interpolate a path as shown in Fig. 2.

The perpendicular distance h and the projection distance d from P_0 to $W_1 W_{i+1}$ is found using the following equations:

$$\underline{V}_1 = \underline{W}_{i+1} - \underline{W}_i$$
$$\underline{V}_2 = \underline{P}_0 - \underline{W}_i$$
$$d = \underline{V}_1 \bullet \underline{V}_2 - |\underline{V}_1|$$

P_0 lies directly above the line segment $W_k W_{k+1}$ if $0 \leq d \leq |\underline{V}_1|$ and the square of the perpendicular distance is given by $h^2 = |\underline{V}_1|^2 - d^2$

This interpolation of GPS signals works fairly well when the remote object follows a path close to the predefined path. However, it may give wrong estimations when the remote object is traveling with great deviation from the predefined path. Though not perfect, the above method does offer a feasible and reasonable solution.

3.2   Adaptive Sampling

The sampling rate of the GPS data is very important. If the sampling rate is too high we may gather a group of very closely located GPS data which means a waste of bandwidth. On the other hand, if the sampling rate is too low, the remote object might have moved too fast that the sampled GPS data will be separated widely apart. This will result in a non-smooth path plotted on the server system. As a result, the sampling rate of the GPS data must be carefully adjusted for the system to work as expected.

In order to make the sampling rate suits different circumstances without readjustment, an adaptive sampling method is introduced. This method will dynamically adjust the sampling rate based on the speed of the remote object being traced. When a GPS data is received, it will be compared with the last sampled GPS data. If the distance between these two positions is greater than the upper threshold value, the sampling rate will be doubled so that more positional data will be sampled. On the other hand, if the distance between these two positions is smaller than the lower threshold value, the sampling rate will then be lowered by half so that the positional data will be sampled less frequently.

4   Data Compression

The GSM mobile network provides a packet data transmission protocol and supports a data rate up to 9600 bps [6]. As a result, it will not be possible to transmit such huge volume of multimedia data from the client to the server without any compression. In our system, the Intel Indeo(R) Video R3.2 codec is employed to compress a video frame of 19 kbytes (160 × 120 8-bit) to around 1 kbytes (160 × 120 24-bit), giving a compression ratio of about 19. The resulting compressed video frame can then be transmitted in about 1 second to the server system. For audio signals, Microsoft Network Audio codec is employed to give a data rate of 1 kb/s (8 kHz, Mono, 8200 baud). The comparison of various compression codecs will be described briefly in the following paragraphs.

4.1   Video Compression

In this study, the compression of the video sequence is achieved by making use of the Installable Compression Manager (ICM). The ICM provides access to the interface used by installable compressors to handle real-time data, and it is the intermediary between the program and the actual compression and decompression drivers. The compression and decompression drivers do the real work of compressing and decompressing individual frames of data.

Table 2. Comparison of video compression drivers

Video Codec	Avg. time (μ sec) for key frame	Avg. size (bytes) for key frame	Avg. time (μ sec) for non-key frame	Avg. size (bytes) for non-key frame
Intel Indeo (R) Video Interactive	239	1314	566	620
Cinepak Codec by Radius	228	1668	32	1416
Intel Indeo (R) Video R3.2	126	1284	56	660
Microsoft Video 1	22	2402	0	6
Uncompressed	0	19200	0	19200

Several video compression drivers have been considered to be used in this project, and their efficiencies are compared in terms of compressed size and compression time. The result is listed in Table 1. All the compression driver are set to have a compression quality of $0\times$ 120 pixels $\times$ 8 bit gray scale. All the tests are performed on the same Intel Pentium-233 MMX computer.

From the above table, it can be seen that Intel Indeo (R) Video R3.2 is well balanced between compressed size and delay and thus is chosen to be used for video compression in this project.

4.2   Audio Compression

In our system, the compression of the audio sequence is done by making use of the Audio Compression Manager (ACM). Likewise, using ACM for compression and decompression has the advantage of being more flexible and the program can choose to use newer and better compression driver when new technology has been developed without re-programming.

Several audio compression drivers have been considered to be used in this project, and their efficiencies are compared in terms of data rate. The result is listed in Table 3.

Table 3. Comparison of audio compression driver

Audio Codec	Freq. (kHz)	Mono/ Stereo	data rate (kB/s)
CCITT A-Law	8	Mono	8
CCITT u-Law	8	Mono	8
Elemedia TM AX24000P music codec	22	Mono	1
GSM 6.10	8	Mono	2
Microsoft ADPCM	8	Mono	4
MSN Audio	8	Mono	1
PCM	8	Mono	8

As shown on the above table, several compression drivers can give a data rate of 1kB/s which is suitable for this project. The one employed in the program is the MSN Audio codec since this codec is commonly available in Windows 95.

5   Experiments and Observations

The whole system had been tested several times with the client system installed on a vehicle traveling along the gray path as shown in Fig. 4.

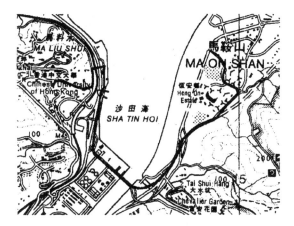

Fig. 4. Test path

5.1 Experimental Results

After connection had been established between the server and client systems, all the GPS data, streaming video data, streaming audio data and still image data could be received from the client system properly. However, the line would be cut when the vehicle moved pass cells of the mobile network. Nevertheless, the client system could automatically recover from the cut-off and no human intervention was needed.

The GPS data received was used to plot the path on the server system, and the interpolation of GPS data of the server system functioned properly for the missing portion of the path being traced. However, the path might not lie exactly on the road of the electronic map, there was a small deviation from the printed roads.

5.2 Analysis and Conclusions

Although in these experiments the connection would be cut-off by the network several times, both the client system and server system could recover from the unexpected cut-off and continue to function. Such cut-off occurs when the vehicle moves from one transmission cell to another in the mobile network. The hand-over of the cells results in temporary cut-off of the line. Such problem is tolerable when the mobile phone is used for voice conversation as human brain can integrate distorted signal, missing signal and noise to recover the original information.

When the signal is too weak, a lot of noise will be resulted which causes error in the data being transmitted. The error correction algorithm of the mobile network might not be able to correct the error in the data when the signal to noise ratio is too low or when the signal is completely lost. This causes the hold-up of the data line and no data can be transmitted though the line is still in connected state. Finally when the error correction algorithm completely fails to recover from the error, the line will be cut off. In such an event, the operator monitoring at the server side should get alerted and make a connection later. In any event, if the path of the mobile object is known ahead, the operator is well prepared to shut down the monitoring system in appropriate locations along the path.

The deviation of the plotted path from the printed roads on the map might be due to the inaccuracy of the GPS receiver. The instability of the GPS receiver caused error in taking the positions of certain sample points for generating the reference coordinates of the map data file. Another possible reason is that a small error in the relative locations of the objects on the map was introduced during the digitization of the printed map into the computer using a desktop scanner.

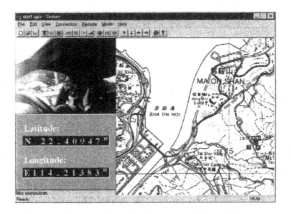

Fig. 5. The server in operation

6   Conclusions

This aim of this project is to demonstrate the feasibility of using a low bandwidth GSM network for transmitting multimedia and GPS positional data in building a real time security system. Since the maximum data rate provided by the GSM network is limited to 9600 bps, the main concern would therefore be how to make good use of the scarce bandwidth.

Our system allows the operator to choose either streaming video, streaming audio or high resolution still image to be transmitted, while the GPS positional data will always be transmitted on a regular basis.

Each transmission frame consists of a header (2 bytes), frame type (1 bytes), data length (2 bytes), actual data plus stuffing characters, checksum (1 byte) and a tail (2 bytes). Assuming about 1% of the actual data are control character need stuffing, the total size of a transmission frame will be 1.1 × data length + 8 bytes.

For still image, 160 × 120 × 8 bits = 19200 bytes (without any lossy compression), the frame size will be 21128 bytes. Its transmission requires 21128 × 8 / 9600 = 17.6 seconds.

For streaming video, each video frame has an average size of 816 bytes after compression. The frame size will be 906 bytes and its transmission requires 906 × 8 / 9600 = 0.76 seconds. Thus the frame rate will be about 1.3 fps, but the actual frame rate would expected to be lower due to the additional time required for compression / decompression and noise during transmission.

For streaming audio, the audio buffer size for 1 second using MSN Audio codec is 1024 bytes. The frame size will be 1134 bytes, requiring 1134 × 8 / 9600 = 0.95 seconds for its transmission. Using this audio codec could theoretically produce a smooth audio output, but the test audio quality was just acceptable as silence gaps existed between samples due to time spent for compression / decompression on software level. Moreover noise in the communication channel reduces the throughput.

Finally, for GPS positional data the data size is 16 bytes. The frame size will be 26 bytes so that 26 × 8 / 9600 = 0.02 seconds are required for transmission. This delay is negligible when compared to other multimedia information.

In spite of the limited bandwidth, our system does demonstrate the feasibility of using the wireless mobile network to transmit separated multimedia data. If new technology in GSM has been developed to provide a data service up to 56k bps, then with current settings, a performance of 6.7 fps with integrated audio data can be achieved. One main drawback of using the wireless mobile network for data transmission is the high air-time cost. The system is thus designed to provide information on demand instead of full-time supervision. When to start and end the supervision and what type of information to be retrieved are completely determined by the system operator, thus the system operator can control the cost of operation.

References

1. Regis J. Bates, Wireless Networked Communications, McGraw-Hill Inc., 1995.
2. Rafael C. Gonzalez, Paul Wintz, Digital Image Processing, Addison Wesley, 1987.
3. B. Hofmann-Wellenof, H. Lichtenegger, J. Collins, GPS Theory and Practice, Springer-Verlag Wien New York, 1994.
4. Tom Logsdon, Understanding the NAVSTAR, GPS, GIS and IVHS, Van Nostrand Reinhold, 1995.
5. Parkinson, Bardford W, Global positioning system: theory and applications, American Inst. Of Aeronautics & Astronautics, 1996.
6. David J. Goodman, Wireless Personal Communications Systems, Addison-Wesley, Jan 1998.

A Witness-Aided Routing Protocol for Mobile Ad-Hoc Networks with Unidirectional Links

Ionuţ D. Aron and Sandeep K. S. Gupta

Department of Computer Science
Colorado State University
Fort Collins, CO, USA
{ionut,gupta}@cs.colostate.edu

Abstract. Mobile ad hoc networks may exhibit unidirectional links due to the nature of wireless communication. Presence of unidirectional links interferes with the control flow of many existing unicast routing protocols for such networks, which adversely effects their performance and limits their applicability. In this paper, we present a new protocol designed to support unicast routing over both bidirectional and unidirectional links in ad hoc networks, while preserving low bandwidth utilization and providing faster and more reliable packet delivery. The WAR (Witness-Aided Routing) protocol is based on the concept of *witness* host, whose role is to help in bypassing a unidirectional or a failed link along the path. We present a preliminary analysis to compare the expected performance of WAR with the Dynamic Source Routing (DSR) protocol.

1 Introduction

Mobile ad hoc networks are being increasingly used for military operations, law enforcement, rescue missions, virtual class rooms, and local area networks. A mobile multi-hop network consists of n mobile hosts (nodes) with unique IDs $1, \ldots, n$. These mobile hosts communicate among each other via a packet radio network. When a node transmits (broadcasts) a message, the nodes in the *coverage area* of the sender can simultaneously receive the message. A node i is called a *neighbor* of node j in the network if node j is in the *coverage area* of node i. This relationship is time varying since the nodes can and do move. At any given time a node i can correctly receive a message from one of its neighbors, say j, iff j is the only neighbor of i transmitting at that time. A multiple access protocol is used by nodes to get contention-free access to the wireless channel.

Radio links are inherently sensitive to noise and transmission power fluctuations. This, as well as problems like the *hidden terminal* [1], can cause temporary or permanent disruption in service at the wireless link level, in one direction or in both. Although intensive research has been done in the last couple of years to design efficient routing protocols for mobile ad hoc networks ([2-9]), there has been little concern about the effect of unidirectional links on the performance of these protocols [10]. The majority of existing protocols assume, either implicitly (e.g. AODV [3]) or explicitly (e.g. TORA [5]), that links are bidirectional. An

H.V. Leong et al. (Eds.), MDA'99, LNCS 1748, pp. 24–33, 1999.

exception, to some extent, is the *Dynamic Source Routing* (DSR) protocol [2], which acknowledges the fact that links could be unidirectional, but does not include any kind of support for such links.

An equally important issue is the way routing protocols cope (or do not cope) with errors in the route. Due to the mobility of hosts and the high instability of links, errors in routing are likely to occur frequently and they can have a significant impact on the performance of a protocol. Little has been done in existing protocols to support recovery from errors. Except for the *Associativity-Based Routing* (ABR) protocol [9], basically no other scheme supports a recovery mechanism to exploit the fact that neighborhoods are probabilistically more stable than individual hosts [1].

The protocol we propose, *Witness-Aided Routing* (WAR), has been designed to support routing over both bidirectional and unidirectional links (Section 2) in mobile ad hoc networks. WAR makes the only assumption that the mobility rate in the network is not as high as to make flooding the only possible routing approach. WAR maintains *low bandwidth utilization*, by collecting routing information on demand, or learning it from transient packets (Section 3.1). In order to minimize the number of routing errors, the protocol includes a *route recovery* component, which, by design, provides also a mechanism for *prioritizing messages* in the network (Section 3.3). WAR is based on the notion of *witness host*, which plays a central role in the routing and recovery process, contributing to faster and more reliable message delivery. (Section 2).

2   Witness Hosts

The *neighborhood* of a mobile host is a union of two sets: the *incoming* and the *outgoing* neighborhood. The *incoming* neighborhood of a host $\mathcal{A}$ is the set of mobile hosts one hop away from $\mathcal{A}$ whose transmissions $\mathcal{A}$ can hear. Similarly, the *outgoing* neighborhood of $\mathcal{A}$ is the set of mobile hosts one hop away from $\mathcal{A}$ which can hear $\mathcal{A}$'s transmissions. Note that if links are bidirectional, the two neighborhoods are the same. It is important to mention here that in WAR mobile hosts do not maintain information about their neighborhoods; this definition is introduced only to facilitate the discussion of the protocol. In what follows, we will refer to the neighborhoods of host $\mathcal{A}$ as $\mathcal{N}_{in}(\mathcal{A})$ and $\mathcal{N}_{out}(\mathcal{A})$.

A *witness* is a host which can overhear a transmission that was not destined to it. Thus, all witness hosts of $\mathcal{A}$ are members of $\mathcal{N}_{out}(\mathcal{A})$. When a witness host is also member of $\mathcal{N}_{in}(\mathcal{A})$ (its transmissions can be heard by $\mathcal{A}$) and:

1. $\mathcal{W} \in \mathcal{N}_{out}(\mathcal{A}) \cap \mathcal{N}_{in}(\mathcal{A})$, and
2. $\mathcal{B} \in \mathcal{N}_{out}(\mathcal{W}) \cap \mathcal{N}_{in}(\mathcal{W})$ or $\mathcal{C} \in \mathcal{N}_{out}(\mathcal{W}) \cap \mathcal{N}_{in}(\mathcal{W})$.

we call it an *active witness* of $\mathcal{A}$ *with respect to a packet* from $\mathcal{A}$ to $\mathcal{B}$ (whose next stop after $\mathcal{B}$ is $\mathcal{C}$). In other words, there is a bidirectional link between $\mathcal{A}$

---

[1] TORA includes a recovery technique, but it is based on link reversal and therefore it is not applicable in networks with unidirectional links

and W and between W and B or W and C. Hence, W can help bypass the link $A \rightarrow B$ (and possibly the link $B \rightarrow C$) in the original route, replacing it with the sub-path $A \rightarrow W \rightarrow B$ (or with $A \rightarrow W \rightarrow C$ respectively). Active witnesses of host A are entitled to act on A's behalf for routing purposes. Henceforth, we will use witness in place of active witnesses when there is no confusion.

An illustration of how witnesses participate in the routing process is shown in Fig.1. Both W_1 and W_2 hear A's transmission to B, which makes them potential active witnesses of A *with respect to the packet* $P_{(A-B)}$ sent to B. At this point, they will wait to see if B attempts to deliver the packet to C, which would mean that B received it from A. If that is the case, their role with respect to the packet $P_{(A-B)}$ reduces to sending an acknowledgement to A (to avoid an error in case A could not hear B's transmission to C). If neither W_1 nor W_2 hear B's transmission to C, they conclude that the packet $P_{(A-B)}$ failed to reach B. In this case, they will both attempt to deliver the packet *directly* to C, although, indirectly, they target B as well. Since W_1 and W_2 do not necessarily have a way to communicate with each other and avoid contention, they will ask C for arbitration before sending the packet. If C rejects their request, it means that it has already received the packet from B and their role reduces to sending the acknowledgement to A. Otherwise, the one selected by C will deliver the packet and then inform A about it. In the example shown in Fig.1, B hears A and

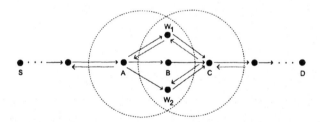

Fig. 1. Hosts W_1 and W_2 witness the transmission from A to B

succeeds in delivering the message to C. But only W_1 can be heard by A, which in fact is enough for the routing step to succeed. If A did not employ the help of its witnesses, and only communicated with B, the result would be a failure, even though the message is traveling towards its final destination. In such a case, A would invoke a *route recovery* protocol (Section 3.3), which would unnecessarily increase network activity and bandwidth consumption.

Formally, a witness host W traverses *two stages* in the routing process of packet $P_{(A-B)}$ from A to B: an initial *passive* stage, followed by an *active* stage. In the passive stage, the witness *listens* to the conversation between A and B and based on its outcome it decides what to do in the active stage. Hence, the active stage may consist of one or two steps: a (potential) *forward* phase and an *acknowledgement* phase. If W hears that B is trying to deliver the packet to $\mathcal{C}$, the forward phase is skipped and $\mathcal{W}$ enters the acknowledgement phase, in which it informs $\mathcal{A}$ that $\mathcal{B}$ received its packet. If not, $\mathcal{W}$ enters the forward phase, attempting the delivery on $\mathcal{A}$'s behalf. If an answer is received back from $\mathcal{C}$ (or from $\mathcal{B}$) to confirm the delivery, $\mathcal{W}$ acknowledges $\mathcal{A}$ and its role ends. The finite state machine describing this process is shown in Fig.2. If no witness

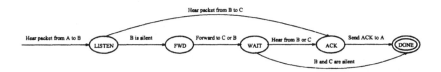

Fig. 2. Witness host state transition with respect to a packet from $\mathcal{A}$ to $\mathcal{B}$

sends a confirmation to $\mathcal{A}$ after a pre-defined period of time, $\mathcal{A}$ will consider that the link to $\mathcal{B}$ failed and will initiate a *route recovery* protocol as described in Section 3.3. There will be cases when the transmission was indeed successful, but nobody could inform node $\mathcal{A}$ about it (i.e. there are no active witnesses). In such cases, since there is no guarantee for $\mathcal{A}$ that the message is traveling towards its destination, the recovery protocol will still be invoked. However, its propagation will be soon terminated by hosts which are aware that $\mathcal{A}$'s transmission was indeed successful (all the witnesses and all the hosts along the original route which have already seen that message).

2.1   Contention Avoidance

A particularly sensitive point of the witness scheme is the potential contention when forwarding a data packet to host $\mathcal{C}$, since they all compete for channel access. In a very dense network this could result in serious delays or even network congestion. In order to reduce the overhead at this point, we use a scheme similar to the channel access scheme used in [12], but modified to allow *one and only one* data delivery. All witness hosts which try to deliver a packet to $\mathcal{C}$, will first ask for permission from $\mathcal{C}$. Upon receiving the request, $\mathcal{C}$ will either reject it (in case it has already received the packet from $\mathcal{B}$ and the witnesses do not know about this) or will only allow one host - say $\mathcal{W}_1$ - to deliver the packet, forcing all the others to remain silent. If $\mathcal{C}$ allows $\mathcal{W}_1$ to transmit, but does not receive the packet after a given time (it is possible that $\mathcal{W}_1$ did not hear the permission), it will select another host from the list of those which made requests and will allow that host to deliver the packet. The process continues until either one of the witnesses successfully delivers the packet, or $\mathcal{C}$ has probed all of them and could not receive the packet from any. Note here that witness hosts will only send transmission requests to $\mathcal{C}$ if they have not heard an implicit acknowledgment from $\mathcal{C}$ to $\mathcal{B}$. Such a *passive acknowledgment* is achieved by operating in *promiscuous receive mode* [2] (i.e. listening to see if $\mathcal{C}$ attempts to forward the packet to the next host along the route).

3   Description of the WAR Protocol

Generally, the existing protocols can be classified in two categories: *reactive* and *proactive*, depending on their reaction to changes in the network topology. Proactive protocols, such as distance-vector protocols (DSDV[4]), are highly sensitive to topology changes. They require mobile hosts to periodically exchange information in order to maintain an accurate image of the network. While convergence is faster in such protocols, the cost in wireless bandwidth required to maintain routing information can be prohibitive. Moreover, mobile hosts are engaged in route construction and maintenance even when they do not need to communicate, so most of the collected routing information is never used. Therefore, many researchers have proposed to use *reactive* protocols, which only trigger route construction or update based on the needs of mobile hosts. Such on-demand protocols, like DSR[2], SSA[6] and AODV[3], have been claimed [13] to outperform the proactive-style protocols. WAR is also a reactive protocol, and has three major components: *route discovery, packet forwarding* and *route recovery*.

3.1   Route Discovery

The route discovery protocol is similar to the one used in DSR[2] in the way the requests are propagated, but it differs in the way the destination host processes them. WAR implements a different request processing in order to *reduce the network traffic* (by limiting the number of route replies) and to *increase the quality of the discovered routes* (by placing route constraints in the request).

The discovery protocol is invoked by a *source host* S every time it needs a route to a *destination host* $\mathcal{D}$, and it does not have one already cached. Host S locally broadcasts a $R_{request}$ message and starts a local timer to decide when to re-send the request in case that no response arrives. All hosts in $\mathcal{N}_{out}(S)$ hear the request and they replicate it, so the $R_{request}$ message propagates in the entire network until it eventually reaches $\mathcal{D}$. Since $R_{request}$ is a broadcast, it is likely that it arrives at the destination on many different paths. The order in which requests arrive is not necessarily an indication about the route's length, so it is not safe to assume that the first request which arrives at $\mathcal{D}$ contains the shortest path. Therefore, S can instruct $\mathcal{D}$ to wait for a defined amount of time for more requests to arrive and return the best route discovered up to that point. This way both the quality of the routes used is improved and the amount of traffic in the network is reduced (compared to the case when *all* requests are replied to immediately). In order to reduce the number of route discoveries, WAR also includes *support for alternate routes*. After sending a route back to S, $\mathcal{D}$ waits for more copies of the request message to arrive and then sends another route. The number of times $\mathcal{D}$ waits for a route, as well as the length of the waiting period are variables which can be set by S in the request message.

There may be cases, like the one illustrated in Fig. 3, when many routes discovered by a request message have a common suffix or prefix. In such cases, it is desirable that only a representative of these routes be returned to S, since otherwise a problem within the common suffix or prefix would break all the available routes. WAR addresses this problem by allowing S to specify a waiting period during which D can filter out such routes, and only return a representative of them. Upon receiving a $R_{request}$ message, a mobile host H checks to see if it

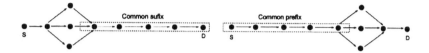

Fig. 3. Situation when more routes share a common suffix or prefix

has seen the message before (by looking up the header in its local history). If so, it drops the message, otherwise it adds it to the history and adds its own ID to the partial route contained in the message. If the host ID is not the same as the final destination of the message (D), H updates the metrics of the partial route contained in the message and verifies if the new metrics satisfy the constraints set by S. If the constraints are satisfied, the message is further delivered to the immediate neighbors of H, otherwise it is dropped. On the other hand, if the host ID is the same as the final destination of the message (D), the host either sends a R_{reply} immediately or waits for more routes to arrive and then sends the best one, according to the instructions set in the CONSTRAINTS field of $R_{request}$ by S.

The structure of a $R_{request}$ message is shown in Fig. 4. It contains a tag (R_REQ) identifying the type of message, the source of the message (SRC), the destination (DST), a sequence number (SEQ_NO), a time-to-live (TTL) indicating the maximum time the message can exist in the network, a partial route (Acc_R) containing all the hosts visited so far, a set of metrics (Acc_M) which will be updated by each host along the way and a set of constraints to guide the decision process as the request proceeds through the network. A R_{reply}

R_REQ	SEQ_NO	TTL	SRC	DST	Constraints	Acc_R	Acc_M

Message header

Fig. 4. The structure of a $R_{request}$ message when it leaves S

message has basically the same structure as a $R_{request}$ message, in which the header has been changed to reflect the direction of the message and its content. Also, if D has a route to S, it includes the route in the header of the reply message so that it does not get disseminated in the entire network. Then, before sending the R_{reply} message, D swaps the SRC and DST fields in the original $R_{request}$ message, updates the tag to R_{reply} and the sequence number according to its own sequence for S. When the R_{reply} message arrives at S, the route it contains (in Acc_R) is entered in the local cache of S, which also sets a lifetime period for the route based on the its metrics. The discovery process ends at this time for S (which can start sending data to $\mathcal{D}$), but routes will continue to arrive until all the $R_{request}$ copies have been received and processed by $\mathcal{D}$.

3.2   Packet Forwarding

The task of the packet forwarding component is not only to ensure delivery of a message along a given path, but to also take corrective actions (invoke the route recovery protocol) when problems in the route are detected. The operation of this component is strictly connected to the witness hosts, discussed in Section 2. WAR uses *source routing* ([2]) in order to deliver packets from a source host S to a destination $\mathcal{D}$. Before sending the packet, S attaches the entire route to it, so at each intermediate host, the packet contains information about the next hop in the route. This eliminates the need for mobile hosts to monitor their direct links by sending periodic *beacons*, which significantly reduces the network overhead. Before a host $\mathcal{A}$ delivers a packet to the next host $\mathcal{B}$ in the route, it removes its own ID from the list of remaining hosts (*to avoid loops*). A packet sent from $\mathcal{A}$ to $\mathcal{B}$ (on a *direct* link) is considered successfully delivered (by $\mathcal{A}$) in two cases:

a) If $\mathcal{A}$ receives a *passive acknowledgement* from $\mathcal{B}$ (it hears $\mathcal{B}$ trying to send the packet to the next host along the route)
b) If $\mathcal{A}$ receives a *positive acknowledgement (ACK)* from any of its witness hosts

Otherwise, $\mathcal{A}$ assumes that the route is broken an initiates a route recovery procedure, as described in Section 3.3.

An equally important role of the packet forwarding protocol is to both supplement the route discovery protocol and to add *route maintenance* capability to WAR *at no extra cost*. When a data packet arrives at a host $\mathcal{H}$, it always contains a route from $\mathcal{H}$ to whatever the final destination of the packet is. Thus, $\mathcal{H}$ can either add that route to its cache (in case none was available before), or replace a lower quality route in the cache. This is done only after $\mathcal{H}$ successfully delivers the packet to the next hop, to avoid adding stale routes into the cache.

3.3   Route Recovery

The time needed to determine new routes in case of an error is critical for a routing protocol. Proactive schemes exhibit a minimal delay, since the routing information is continuously updated. On the other hand, on-demand protocols can experience considerable delays when an error occurs and a new route is needed. Some existing protocols, like DSR[2], address this problem by notifying the sender about the error. This is inefficient for two reasons. First, errors may appear in the vicinity of the final destination $\mathcal{D}$ (and far away from the sender S). Second, the bandwidth used for sending a negative acknowledgement and for re-sending the packet from S could actually be used for sending a recovery request in the vicinity of the host in error ($\mathcal{H}$). By involving an extended neighborhood of

$\mathcal{H}$ in the recovery process, WAR attempts to quickly and inexpensively bridge the gap created in the route. This would allow the packet to travel to its destination, as opposed to delaying it until S finds another route.

WAR implements error handling at *two levels*. First, witness hosts offer support for packets to bypass broken or unidirectional links (*error prevention*). If witness hosts fail to acknowledge success to their *witnessed* host ($\mathcal{H}$), the second level, *error recovery*, is activated. Host $\mathcal{H}$ broadcasts a copy of the original message, with the tag changed to $R_{recovery}$, to its neighbors. Before broadcasting the message, $\mathcal{H}$ increments the PROBLEM counter present in the message (which was set to 0 by S when the original message departed). This *route maintenance* step helps S find out about the problems with its route when it receives the final acknowledgement ACK from $\mathcal{D}$ (which piggybacks the PROBLEM counter on ACK). After sending out the recovery message, $\mathcal{H}$ will drop the original message, and its role in the routing process ends. No acknowledgment is necessary for recovered messages. As soon as one of the hosts in the remaining route (indicated in the message header) is reached, the message tag is changed back into *DATA* and it continues its travel as a normal data packet. If the recovery succeeds, the final destination $\mathcal{D}$ will eventually inform S, otherwise S times out and re-sends the packet.

The number of steps a message can travel as a route recovery message (with a $R_{recovery}$ tag) is indicated in the constraint field attached to the original data packet by the sender (the Recovery Depth value). When the Recovery Depth counter becomes zero (being decremented by each host which receives the $R_{recovery}$ message), the message is no longer propagated and the recovery fails (on that branch of the network). This way, WAR also provides a framework for setting *message priorities*. A greater Recovery Depth will cause the recovery protocol to be more insistent, increasing the chances of success (at the expense of bandwidth consumption).

4   Preliminary Performance Analysis

We next model the effect of route correction and compare it with DSR, which does not use a recovery mechanism. Let $F_{\mathcal{X}}[m]$ be the probability of failure in delivering a message by Protocol $\mathcal{X}$, where $\mathcal{X}$ is either WAR or DSR, on a path of length m. Let p be the steady state probability that a message is delivered to the next hop along the route. In case of WAR, let q be the steady state probability that the route is corrected using active witness nodes in case of failure. Further, for sake of simplicity we assume that m is an even number. Hence, for DSR we have:

$$F_{\text{DSR}}[m] = 1 - p^m, \tag{1}$$

since p^m is the probability that the message will be delivered along a path of length m. In case of WAR, we have

$$F_{\text{WAR}}[m] = 1 - \left(p^2 + \left(1 - p^2\right) q\right)^{m/2}, \tag{2}$$

since the probability that a message is delivered two hops away is $p^2 + (1 - p^2)q$ and, consequently, the probability that the message is delivered to the destination m hops away from the source is $(p^2 + (1 - p^2)q)^{(m/2)}$. We can draw following conclusions from this simple modeling. For both the protocols as m grows larger the probability of failure increases. However, WAR has lower probability of failure than DSR for same value of route length m. Further, as q approaches 1, $F_{\text{WAR}}[m]$ approaches 0; and hence becomes increasingly independent of p. Hence, by applying intermediate route correction we can decrease the probability of failure considerably. It is easy to determine that the average number of transmissions[2] required in DSR to deliver a message at a distance of m is $\frac{1}{p^m}$ and that for WAR is $\frac{1}{(p^2+(1-p^2)q)^{(m/2)}}$. Since $\frac{1}{p^m} \geq \frac{1}{(p^2+(1-p^2)q)^{m/2}}$, WAR requires fewer number of tries to deliver a message than DSR. However, it should be noted that more bandwidth is required to deliver a message in WAR than DSR. Let on an average WAR require α times more bandwidth per hop than DSR. Then, WAR is more bandwidth-efficient than DSR if

$$\frac{m\alpha}{\left(p^2 + \left(1 - p^2\right)q\right)^{m/2}} < \frac{m}{p^m}, \qquad\qquad m \geq 1$$

$$\alpha < \left(1 + \frac{\left(1 - p^2\right)q}{p^2}\right)^{m/2} \tag{3}$$

Note that $\left(1 + \frac{(1-p^2)q}{p^2}\right) \geq 1$ in the right hand side of Eq. 3. Therefore α should be less than $\left(1 + \frac{(1-p^2)q}{p^2}\right)$, if WAR is to be more bandwidth efficient than DSR for all values of m. We expect that α would be close to 1 since WAR uses atmost k extra one-hop explicit ACK messages, where k is number of active neighbors of a node, and the size of ACK packets is much smaller than DATA packets. Consequently, it is likely that Eq. 3 would be satisfied for wide ranges of p and q. For example, assuming $p = 0.75$ and $q = 0.5$ we require $\alpha \leq 1.39$ for WAR to perform better than DSR.

5   Conclusions

The main goal of the WAR protocol is to provide an efficient and low cost routing solution for mobile ad hoc networks in the presence of *unidirectional links*. By using witness hosts, WAR is designed to *reduce the number of transmission errors*, implicitly reducing the communication delay and the overall bandwidth consumption. The role of the recovery mechanism is to reduce *the effect* of errors and to provide a framework for *prioritizing messages* in the network. *Route maintenance* is done with no extra expense in terms of bandwidth, by learning routes from transient packets and by using the recovery mechanism to collect information about the status of the routes being used. The *on-demand* character

---

[2] The number of times a source node has to send the packet to the destination node before the packet gets delivered to the destination node.

of the route discovery protocol eliminates the need for periodic advertisements, reducing considerably the bandwidth utilization and power consumption.

WAR's most important advantage is that it does not rely on the existence of bidirectional links, which brings it closer to the reality in mobile ad hoc networks. Our preliminary performance analysis shows that WAR will perform better than DSR with almost the same resource utilization. In the future, we plan to test its robustness and compare its performance with other routing protocols through simulation, under different network conditions.

References

1. A. Tanenbaum. *Computer Networks*. Prentice Hall, Upper Saddle River, NJ, 3rd edition, 1996.
2. D. Johnson and D. Matlz. Dynamic source routing in ad hoc wireless networks. *Mobile computing*, pages 153–181, 1996.
3. C. E. Perkins and E. M. Royer. Ad-Hoc on-demand distance vector routing (AODV). *IEEE*, 1999.
4. C. Perkins and P. Bhagwat. Highly dynamic destination-sequenced distance-vector routing (DSDV) for mobile computers. *ACM SIGCOMM'94 Conference on Communications Architectures, Protocols and Applications*, pages 233–244, Aug 1994.
5. V. D. Park and M. S. Corson. A highly adaptive distributed routing algorithm for mobile wireless networks. *Proceedings of INFOCOM '97*, April 7-11 1997. Kobe, Japan.
6. R. Dube, C. Rais, K.-Y. Wang, and S. Tripathi. Signal stability based adaptive routing (SSA) for ad-hoc mobile networks. *IEEE Personal Communications*, Feb 1997.
7. Y.-B. Ko and N. H. Vaidya. Location-aided routing (LAR) in mobile ad hoc networks. Technical Report 98-012, Texas A&M university, Department of Computer Science, College Station, TX 77843-3112, June 1998.
8. Z. J. Haas and M. R. Pearlman. The zone routing protocol (ZRP) for ad-hoc networks. *Internet draft*, November 1997.
9. C-K Toh. *The Cambridge Ad-Hoc Mobile Routing Protocol*. Kluwer Academic Publishers, wireless atm and ad-hoc networks: protocols and architecture - second printing edition, 1997.
10. R. Prakash and M. Singal. Unidirectional links prove costly in wireless ad-hoc networks. *DIMACS Workshop on Mobile and Wireless Networking*, 1999. To appear.
11. E. M. Royer and C.-K. Toh. A review of current routing protocols for ad hoc mobile wireless networks. *IEEE Personal Comunications*, April 1999. alternative http://beta.ece.ucsb.edu/ eroyer/research.html.
12. V. Bharghavan, A. Demers, S. Shenker, and L. Zhang. MACAW: A media access protocol for wireless LANs. *Proceedings of ACM SIGCOMM '94*, pages 212–25, 1994.
13. J. Broch, D. A. Maltz, D. B. Johnson, Y.-C. Hu, and J. Yetcheva. A performance comparison of multi-hop wireless ad hoc network routing protocols. *MOBICOM98*, 1998. Dallas, Texas.

Assessing Opportunities for Broadband Optical Wireless Local Loops in an Unbundled Access Network

Edward Mutafungwa[1] and Kamugisha Kazaura[2]

[1] Communications Laboratory, Helsinki University of Technology, P.O. Box 2300, FIN-02015 HUT, Finland
edward.mutafungwa@hut.fi
[2] Global Information and Telecommunications Institute, Waseda University, 55-S-602 3-4-1 Okubu, Shinjuku-ku Tokyo 163-8555, Japan
kazaura@witl.rise.waseda.ac.jp

Abstract. Optical wireless local loops (OWLL) based on infrared wireless communication links, are class of wireless access networks capable of delivering the type high-bandwidth services that are becoming increasingly popular with business and residential customers. By combining the use of advanced optical transmission techniques with the flexibility of wireless communication, we describe the possibilities that are created by utilizing this technology in the access network. The possible implementation options for deploying OWLL systems and a performance analysis of a typical OWLL link is also presented. Additionally, we provide a comparison with the other wireless access technologies based on attributes that promote the competitiveness of a particular access network technology.

1 Introduction

The deregulation of the telecommunications industry has created a competitive environment in the access network or the "last mile" part of the telecommunications infrastructure. As a consequence, competing operators have had to develop cost effective ways of bringing broadband access to the customers premises and enable the convergence of voice, data and video services [1]. The pace of these changes is expected to increase due to the volume deployment of future high-speed data services and other applications with an insatiable bandwidth demand. The access network has traditionally been dominated by legacy equipment based on twisted-pair copper cables, but fixed wireless access techniques are rapidly emerging as the technology of choice in the unbundled access networks. They ensure that the investment costs are minimized, hence making the service affordable to all classes of customer. It is also possible to reduce the installation time, introduce a limited degree of mobility into the access network and only install terminal equipment when the customer requests the service. A typical wireless local loop (WLL) system consists of a wireless subscriber unit (WSU), operation and management center (OMC), switching unit and an access network unit

H.V. Leong et al. (Eds.), MDA'99, LNCS 1748, pp. 34–44, 1999.

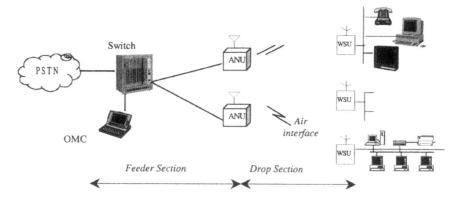

Fig. 1. Generalized wireless local loop structure

(ANU) (Fig. 1). The latter unit may house a WLL controller, access manager (AM), home location register (HLR) and the transceiver unit.

The limited capacity of wireless radio links continues to be a major disadvantage of wireless networks when compared to wireline access networks. This situation is likely to remain the same in the foreseeable future due to the introduction of novel wireline technologies such as passive optical networks, high-speed digital service lines (xDSL) or cable modems. In this paper we assess the opportunities created by using optical wireless local loop (OWLL) techniques capable of delivering two-way fiber-like data rates whilst maintaining most of the advantages wireless techniques.

2   Optical Wireless Systems

This decade has seen sudden surge in the R&D activities and initiatives for optical wireless systems [4, 5]. The establishment of the Infrared Data Association (IrDA) by a group of leading companies and the subsequent standardization activities, is one of the notable outcomes of these activities. The essence of optical wireless system is lays in the fact that light (operating in the infrared or IR band of the EM spectrum) propagates through air, instead of being confined in fiber waveguides as is the case in optical communications systems. This eliminates the costly (and demanding) task of fiber cabling and takes advantage of the absence of optical spectrum licensing requirements.

So far most of the work done in this area has focused on the development of infrared indoor communication systems, where the coverage is confined within a single room, with the communicating terminals within meters of each other. A range of optical wireless products currently available on the market, for basic tasks suitable for small-office-home-office (SOHO) demands such as the connection of portable computers, copiers, facsimile machines, printers, scanners etc., provision of camera-monitor connectivity in CCTV security applications and infrared wireless LAN applications. But the increasing maturity of optical components has enhanced the feasibility outdoor optical wireless links, with results of some field trials indicating error free transmission for distances in excess of 1 km. The most notable of these trials to date managed to achieve an error free transmission of a 10 Gbit/s light signal (four STM-16 wavelength division multiplexed channels) over a 4.4 km link [6]. Such transmission distances can easily cover most drop section span lengths (e.g. the average span length in the Helsinki metropolitan area is 1.8 km).

3 Possible Owll Implementation

3.1 Link Designs

Several optical wireless link designs are available and these could be classified according to their degree of directionality and/or the existence of a line-of-sight (LOS) path (Fig. 2). The choice of a particular link design will depend on, among other factors, the power efficiency required, link distance, presence of obstacles in the propagation path and type of communication (e.g. point-to-point, point-to-multipoint).

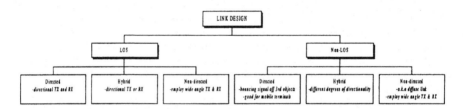

Fig. 2. Classification of optical wireless link designs

In practice, a typical service area will have complex demographic patterns [1] influences the link choice for OWLL design. Previous power budget analysis and channel capacity evaluations have identified directed or hybrid LOS links as being best suited for such outdoor applications [4]. The overall power budget of an OWLL link mainly dependant on the atmospheric-induced loss incurred along the propagation path, with the primary contributions being [5]:

- *free space loss* – proportion of transmitted optical power that is incident on the receiver,
- *air absorption loss* – due to presence of ionic impurities in the air that create a vibration band in the near infrared region,
- *scattering losses* and *beam refraction* – attributed to the attenuation caused by water droplets in rainy, misty, snowy or foggy conditions,
- *scintillation losses* – this is a form of scattering loss caused by the non-uniform refractive index (due to solar heating) of different pockets of air, this can lead to optical power level fluctuations of up to 30 dB.

Furthermore, the presence of ambient light from external light sources such as fluorescent street lamps or the sun, impair the transmitted signal by introducing photonic interference noise at receivers with a wide field of view (FOV). These stray light signals can be eliminated using plastic infrared filters or employing robust receiver designs.

3.2   Modulation and Demodulation Options

The recommended modulation and/or encoding scheme for low-cost optical wireless systems with a sufficient signal power is on-off keying (OOK) based on intensity modulation [1] (IM). the direct-detection[2] (DD) down-conversion technique is used whereby (an operation that would have been difficult to perform using amplitude, phase or frequency modulation. For OWLL applications, vertical-cavity surface-emitting laser diodes (LD) operating in the Class 3B band are preferred as they offer high electro-optic conversion efficiencies (30–70%), very narrow spectral widths and multi-gigabit bandwidths (this at least an order of magnitude more than light emitting diodes). To meet eye safety requirements due to the use of high powered LD sources, the communicating equipment are located at rooftop levels or high towers where humans are unlikely to cross the beam. Alternatively, the beam is passed through holograms thus transforming the LD source into a Class 1 eye safe device according to the IEC standard's allowable exposure limits [4]. Avalanche photodiodes could ideally provide a high enough SNR when used in DD receivers. But since they have temperature-dependent gain, high costs and operating voltages, PIN diodes are preferred instead.

Other modulation techniques that could be considered for OWLL include the pulse position modulation (PPM) and digital pulse interval modulation (DPIM) techniques. The former offers higher average power efficiency than OOK but at penalty of less bandwidth and increased system complexity as a result of its intricate encoding and synchronization procedures. DPIM avoids such bandwidth and complexity problems but suffers from long transmission times, causing delays that might be deemed unacceptable for some services [7].

3.3   Multiple Access Techniques

Since light beams cannot penetrate opaque objects, the efficiency in the IR bandwidth reuse is improved drastically. The SNR of a DD receiver has a linear relationship with the received optical power, thus it is never possible to increase transmitter power (or decrease noise) until the system becomes interference limited. Time division multiple access (TDMA) is presently the popular choice for IR wireless LANs, different users share the same optical channel and electrical multiplexing techniques enable simultaneous transmission. But these electrical methods create a transmission speed bottleneck that might be detrimental to

---

[1] The desired waveform is modulated onto the instantaneous power of the carrier

[2] The photodetector produces a current that is proportional to the received instantaneous power

the OWLL in the long run. There are also some cost concerns brought about by the need for complex transmitter coordination. The use optical CDMA-based systems [8] offer a relatively better cost and bandwidth efficiency performance as well as increased transmission security, but they are still at an early stage of development. In subcarrier multiple access (SCMA) systems, different users transmit simultaneously at different subcarrier microwave frequencies that are then multiplexed onto an optical carrier. This offers significant cost reductions and a possibility of using mature microwave circuit technology. But the throughput is limited since in practice different channels must reside within the bandwidth of a single optical carrier.

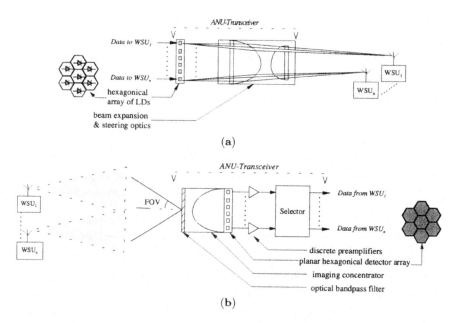

Fig. 3. A SDMA-based optical wireless local loop (a) downlink and (b) uplink structure

Space division multiple access (SDMA) implementation is based on hybrid LOS links using an angle-diversity receiver at the ANU-transceiver to detect signals of the same wavelength band emanating from proximal geographical locations. This offers a relatively better performance (in terms of ambient noise rejection & co-channel interference reduction) compared to a single-element receiver, but at the price of increased cost, receiver employs an imaging concentrator to significantly simplify the receiver structure and support even more receiving elements [5]. A hexagonal array of LD proceeded by lenses for beam expansion and steering may be used to provide a spatially addressable downlink communications (Fig. 3a). In turn, the WSU need only be equipped with single-element receivers and transmitters, thus minimizing cost & power require-

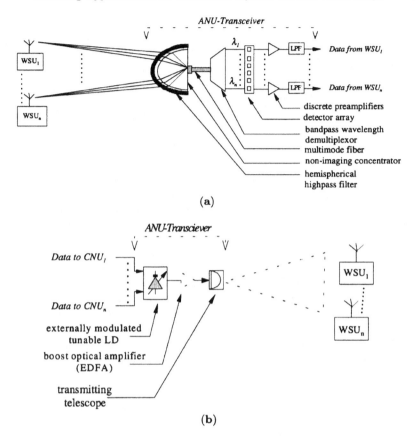

Fig. 4. A WDMA-based optical wireless local loop (**a**) uplink and (**b**) downlink structure

ments. The uplink design will be based on a similar spatial addressing technique (Fig. 3b) [9].

In the case of wavelength division multiple access (WDMA) each WSU is accessed by a signal of pre-designated wavelength channel. This eliminates the need for transmitter arrays at the ANU, using instead a single non-directed tunable LD to simultaneously produce light at N different wavelengths $\{\lambda_i : 1 \leq i \leq N\}$. Single-element tunable receivers are not considered because their input filter designs limit the FOV and are difficult to fabricate. Therefore the ANU multi-wavelength receiver can be implemented using a hemispherical high pass filter and a telescoping non-imaging concentrator to couple light into a multimode fiber (singlemode fibers are too narrow to collect to this beam). The output of the fiber is eventually fed into a wavelength demultiplexor which separates the individual signal channels contained in the aggregate WDM signal and these signals are eventually passed on to a bank of PIN receivers.

3.4   Performance Analysis of LOS OWLL Links

In this section, we analyze of the performance of LOS links that could be used in SDMA-based OWLL systems. An optical signal would normally to be incident on more than one photodetectors on the detector array. A weighted sum of the photocurrents emanating from N photodetectors would then give the received electrical signal. The SNR obtained by this technique is given by [11]:

$$SNR = \sum_{i=1}^{N} \frac{\Re^2 P_{RX,j}^2}{\sigma_i^2} \tag{1}$$

where $\Re$ is the detector responsivity; $P_{RX,i}$ and σ_i are the respective average detected signal and noise power at the ith photodetector. Assuming that the incoming beam subtends an angle ϕ with respect photodetector surface normal and if this angle is less than the reciever acceptance angle ϕ_a then the received signal power is

$$P_{RX,i} = I_{TX} e^{-\alpha L} \frac{A}{L^2} T_F T_{C,\phi} F_{\phi,i} \cos \phi \ . \tag{2}$$

where I_{TX} is the transmitter's radiant intensity, α is the atmospheric attenuation coefficient, $F_{\phi,i}$ is the fraction of power received by the photodetector, A is the detector area, L is the link distance, T_F and $T_{C,i}$ are the filter and concetrator transmission factors respectively.

The detected noise power is a sum of several independent noise sources that are approximately Gaussian in their statistics. The noise power at the ith photodetector can be estimated by [11]:

$$\begin{aligned}
\sigma_i^2 = \ & 8q\Re\pi A N_A \Delta\lambda T_F T_{C,\phi} K_1 B \cos\phi \sin^2\left(\frac{\phi_a}{2}\right) \\
& + \left(\frac{K_2}{R_L} + \frac{(2\pi CB)^2}{G} K_3\right) 4k_{\mathrm{B}} T B \ .
\end{aligned} \tag{3}$$

The first term represents the shot noise resulting from the background ambient noise described by the power spectral density N_A, with q being the electronic charge, $\Delta\lambda$ is the noise equivalent bandwidth of the receiver circuit, B the bit rate and $K_1 = 0.6$. The second term represents the aggregate thermal noise associated with the photodetector circuitry. The symbol R_L represents the load resistance, C is the preamplifier's input capacitance, G is the preamplifier transconductance, k_{B} is Boltzmann's constant, T is the operating temperature with K_2 and K_3 being 0.562 and 0.13 respectively.

Assuming that zeros and ones are equally likely to occur, a zero extinction ratio and neglecting the effects of ISI, the BER is given by [12]:

$$BER = \frac{1}{2}\left[1 - erf\left(\frac{SNR}{2\sqrt{2}}\right)\right], \tag{4}$$

where

$$erf(x) = \frac{2}{\sqrt{\pi}} \int_0^x e^{-y^2} dy \ . \tag{5}$$

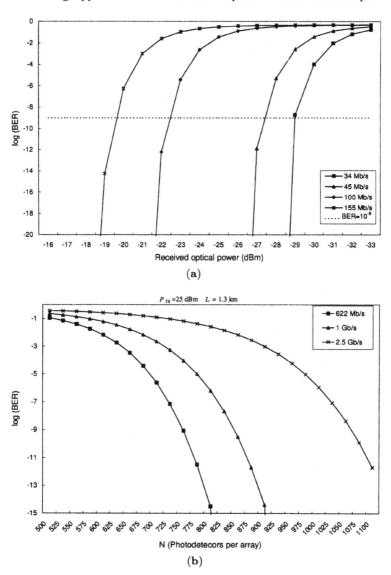

Fig. 5. (a) BER curves for a 1.3-km OWLL loop link using imaging receivers (b) link performance using a large number of individual photodiodes

As a numerical example we use the following values: $C = 11\,\text{pF}$, $T = 300\,\text{K}$, $N_A = 43\,\text{mW}/(\text{m}^2.\text{nm})$, $\Delta\lambda = 80\,\text{nm}$, $R_L = 50\,\Omega$, $G = 0.03\,\Omega^{-1}$, $\Re = 0.56\,\text{A}/\text{W}$, $L = 1.3\,\text{km}$ and $\alpha = 17\,\text{dB}/\text{km}$ (measured average in rainstorm or moderate snowy conditions [13]).

Figure 5a shows a variation BER with the received optical power levels for four different data rates and 100 photodetectors in the receiver array. The excess losses due to coupling at the transmitter and the receiver mode mismatch have been ignored in this analysis. By increasing the number of number of photodetectors, error-free transmission higher bit rates is possible as long as the area incident beam doesn't exceed the area of a single photodetector (Fig. 5b). This places an upperbound on the maximum link capacity, this can be extended but at the penalty of increased ambient and thermal noise.

Similar analysis could be carried out for WDMA-based OWLL by replacing the parameters for imaging receivers with equivalent parameters for single-element receivers. The effects adjacent channel crosstalk and spontaneous optical amplification noise could be modeled for a more accurate sensitivity estimation.

4   Comparison with Other WLL Systems

For OWLL to be competitive they have to offer some significant advantages over the more established radio based WLL technologies [3]. These include:

i. *FWAS*: These are based on cordless mobile radio standards (e.g., CT-2, DECT) or proprietary technologies (e.g., Lucent's Airloop, Nortel's Proximity series).

ii. *MDS*: The microwave distribution service could be either the multichannel multipoint distribution service (a.k.a. the 'wireless cable') or the local multipoint distribution system (LMDS). These systems have proved to be particularly suitable for video broadcasting in areas with rugged terrain.

iii. *CBS*: Cellular-based systems that offer both fixed wireless and mobile services from the widely deployed cellular infrastructure. Originally designed for cellular voice communication and high-tier coverage.

iv. *SATS*: The satellite systems that make use of a constellation of low-earth-orbiting (LEO) satellites (e.g. recently launched Iridium system [10]), geostationary-earth-orbiting (GEO) satellites (e.g. Astrolink by Lockheed Martin) or hybrid-earth-orbiting (HEO) satellites (e.g. Star Lynx by Hughes).

The overview and description of the cost and performance limiting attributes of the various WLL technologies (Table 1) offers a level playing field for the comparison of the relative merits and drawbacks of the respective technologies.

5   Conclusions

In this paper an assessment of the apparent potential of OWLL systems was carried-out. By combining high information carrying capacity and benefits of

Table 1. Attributes of various WLL systems (Shading denotes a strong advantage)

Attributes	OWLL	FWAS	MDS	CBS	SATS
Operating frequencies	1-250 THz	3.4-3.6 GHz	28-38GHz,2-3 GHz	800-900MHz, 1.5GHz,2GHz	ka-, ku- & V-bands
Frequency licensing?	None	Yes	Yes	Yes	Yes
Data rate (downstream)	≤ 2.5Gb/s per channel	Up to 128 kb/s, future 25 Mb/s	~100s Mb/s	2Mb/s (UMTS)	100s kb/s
Possible range	≤ 5 km	Up to 35 km	< 8 km	≤ 15 km	Unlimited
Multipath fading?	None	Yes	Yes	Yes	Yes
Relative user equipment costs	High	Lower	Low	Lower	Medium
LOS requirement	Stringent	Relaxed	Stringent	Relaxed	Relaxed
QoS/availability	Weather dependent	Reasonable	Weather dependant	POTS quality	(~99.95%)
Service offerings	Unlimited	Limited	Unlimited	Limited	Limited
Service type	Symmetrical	Asymmetric	Symmetrical	Asymmetric	Asymmetric
Standards	IrDA standards	No agreed Standards	No agreed standards	No agreed standards	No agreed standards

wireless systems, this technology offers qualities that are suited to the demands of both customers and incumbent operators. Judging from the current traffic growth trends, this solution is notably future proof by virtue of its high capacity and symmetrical capabilities. With the current cost wiring a residential area with fiber optic cables averaging about US$ 5000 per home passed, OWLL is cheap alternative that might enable the deployment of a virtual fiber-to-the-home systems.

The current improvements in manufacturing techniques of optical components and their increased volume production with time will lead to significant equipment price drops. This should place this technology in a price-bracket that will prove affordable to both business and residential customers. Furthermore, novel installation techniques should help maintain continuos LOS link against any future obstructions such as new buildings. All these factors taken together should fulfil the OWLL's potential as a major access network technology.

References

1. Ims L. (ed.): Broadband Access Networks. 1st edn. Chapman & Hall, London, (1998)
2. Webb, W.: Introduction to Wireless Local Loops. 1st edn. Artech- House, Boston (1998)
3. Noerpel, A., Lin, Y., "Wireless Local Loop: Architecture, Technologies and Services," IEEE Personal Comm. Mag. 5 (1998) 74–80
4. Heatley, D., Wisely, D., Neild, I, Cochrane, P.: Optical Wireless: The Story So Far. IEEE Comm. Mag. 36 (1998) 72–82
5. Kahn J., Barry J.: Wireless Infrared Communications. Proc. of IEEE. 85 (1997) 263–298

6. Nykolak, G., Szajowski, P.F., Jacques, J., Presby, H.M., Abate, J. A., Tourgee, G. E., Auborn, J.J.: 4x2.5 WDM Free-Space Optical Link at 1550 nm. OFC'99 Technical Digest. Post-deadline Paper. (1998) 11/1-3

7. Ghassemlooy, Z., Hayes, A., Seed, N., Kaluaurachchi, E.: Digital Pulse Interval Modulation for Optical Communications. IEEE Comm. Mag. 36 (1998) 95–99

8. Andonovic I., Huang, W.: Optical Code Division Multiple Access Networks. Proc. of PHOTONICS'98. 1 (1998) 29–34

9. Street, A., Samaras, K., O'Brien D., Edwards, D.: High Speed Wireless IR-LANs Using Spatial Addressing. Proc. of PIMRC '97.3 (1997) 969–973

10. Fossa, C., Raines, R., Gunsch, G., Temple, M.: An Overview of the Iridium Low Earth Orbit (LEO) Satellite System. Proc. of IEEE Conf. on Aerospace & Electronics. (1998) 152–159

11. Djahani, P., Kahn, J.: Analysis of Infrared Wireless Links Employing Multi-Beam Transmitters and Imaging Diversity Receivers. Submitted to IEEE Trans. on Comm., (1998)

12. Keiser, G.: Optical Fiber Communications. 2nd edn. McGraw-Hill New York (1993)

13. JOLT Ltd.: Long Range High Speed Wireless Link. Technical Specifications TS-U4000 (1999)

On Simulation Modeling of Information Dissemination Systems in Mobile Environments

Wang-Chien Lee, Johnson Lee, and Karen Huff

GTE Laboratories Incorporated, 40 Sylvan Road, Waltham, MA 02451, USA
{wlee, jlee, khuff}@gte.com

Abstract. In this paper, we propose dynamic discrete-event based models for information dissemination systems in both single cell and multiple cell mobile environments. We demonstrate that this novel approach can flexibly model channel borrowing and hand-off scenarios. Finally, we implement two simulation systems based on the single cell model. Our simulations show that broadcast channels can effectively alleviate traffic overload in a cell.

1 Introduction

In the forthcoming era of ubiquitous information services, wireless bandwidth is one of the most valuable resources. How to utilize the limited wireless bandwidth to provide a wide range of services for mobile users is a critical research issue.

Cell splitting and *information broadcasting* are two of the techniques for alleviating demands for wireless bandwidth. Cell splitting is to divide a large cell into a number of smaller cells. By reducing the size of the cells, more cells per area will be available. Thus, the number of channels and traffic capability in the area is increased. On the other hand, information broadcasting is very useful for disseminating data of common interest, e.g., stock ticks and breaking news, to a large population of mobile users. By dedicating some channels for broadcasting popular information, overloaded traffic on traditional point-to-point, on-demand channels may be alleviated. Moreover, broadcast channels are a natural approach for realization of *push*-oriented information dissemination which balances *pull*-based data access on point-to-point channels. As a result, research and new developments in mobile information systems have involved both broadcast channels and point-to-point channels as communication media for wireless information dissemination systems [IV94,AAFZ95,LHL99,SRB97]. In this paper, we consider performance modeling of information dissemination systems in single and multiple cell mobile environments.

An interesting and important aspect of mobile computing in multiple cell environments is *roaming*. In order to facilitate wireless roaming, base stations have to negotiate with each other to ensure continuous service to a user moving across cells. Whenever a mobile user moves into a new cell, his mobile computer registers with the base station in charge of that cell. This registration process usually involves communication between the old base station and the new base station to *hand-off* information associated with the mobile user and computer.

H.V. Leong et al. (Eds.), MDA'99, LNCS 1748, pp. 45–57, 1999.

During off-peak hours, base stations usually have light competition for channels and thus can ensure a smooth hand-off for the roaming users. During peak hours, however, base stations may have too many mobile users waiting for channels to be available. In this situation, the base station may adopt a policy to assign higher priorities to the already connected, roaming users with a hope that channels can be allocated for the roaming users before they enter the cell. Research on supporting communication hand-off has been presented in [BB95,TJ91].

In addition to the issue with hand-off, *channel borrowing* is another interesting aspect of mobile computing which appears in multiple cell environments. Basically, wireless channels are allocated to the cells based on geographic distribution of the traffic load. If some cells become more overloaded than others, it may be possible to reallocate channels by transferring frequencies from the lightly loaded cells to the heavily loaded cells. This is an interim solution, because the borrowed frequencies (and channels) have to be returned to the original cells when the traffic in the original cells grows.

Performance analysis of the mobile computing systems is a critical task for planning and operation of the systems. Queueing theory based models for performance analysis of the mobile systems have appeared in the literature [LHL99,IB94], mostly in the context of single cell mobile computing environments. These queueing models, assuming Poison and Exponential distributions for inputs and service time, are not faithful to the dynamic real world situations. Moreover, these models do not provide global analysis of systems involving cross-cell activities. With increasing offerings of ubiquitous services and demands for wireless bandwidth, performance analysis and global optimization of systems and services in multiple cell mobile environments is critical for success in today's competitive market. In this paper, we propose a simulation model, based on dynamic discrete-event graphs, for a wireless information dissemination system in both the single cell and multiple cell mobile environments.

The rest of the paper is organized as follows. In Section 2, we propose a simulation-based performance analysis model for the information dissemination system in the single cell environment. In Section 3, we extend the model to cover multiple cell mobile computing scenarios, i.e., channel borrowing and hand-off. In Section 4, we describe our implementations and discuss simulation results. Finally, Section 5 concludes the paper.

2   Dynamic Discrete-Event Modeling

Queueing theory based models have been used for performance analysis of the mobile systems. In [LHL99], for example, queueing models and analytical methods are used for performance evaluation of the point-to-point communication channels and the broadcast channels, respectively. Overall system performance is then obtained by combining weighted results from both channels. One major defect of this approach is the lack of one coherent analytical model for the whole system. To remedy the above problem, we develop in this paper a new simulation model for wireless information dissemination based on *dynamic discrete-event modeling* technique [HP89].

Dynamic discrete-event modeling is a simulation modeling technique broadly used in experiment-intensive fields such as Nuclear Physics. Compared to queueing theory based models and simulations, the dynamic discrete-event based model and simulations provides us the following important advantages:

- The logical correctness of the model is easy to verify;
- The simulations are easy to implement;
- The simulations are more realistic since some constraints in the queueing model are relaxed:
  - Queueing models assume an infinite duration for the simulation;
  - Queueing models assume either Poison or Exponential distributions for parameters in the model;
  - Queueing models only handle steady states.

| Event i | Unconditional Transition | Conditional Transition | |
| (a) | | | (b) |

Fig. 1. Event graph: (a) basic notations; (b) conditional event transition

Basically, a discrete-event simulation model can be viewed as a model of the interaction of discrete events occurring in the system and the system's state variables. These interactions can be represented as an event graph where the nodes represent the events and the arrows represents transitions between two events. In the graph, a transition between two events can be unconditional or conditional. The time delay and condition for a transition are also denoted in the graph. Figure 1(a) illustrates the basic notations of the graph and Figure 1(b) implies that providing the necessary condition C1, event i will lead to event j with a delay of t.

In Figure 2, we present an event graph for the wireless information dissemination system in single cell environment. Descriptions of events, conditions and elapsed time in the graph are also given in the following.

EVENTS

    E1: a user attempt to connect to the base station.

    E2: blocked user re-send connection request to the base station.

    E3: a point-to-point channel allocated and service session begins.

    E4: session ends.

    E5: user tunes into the broadcast channels.

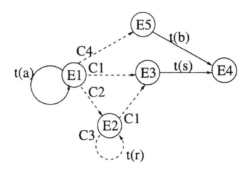

Fig. 2. Single Cell

CONDITIONS

C1: $m < c$.

C2: $m = c$.

C3: restless user keeps re-sending connection requests.

C4: user has valid access information for the broadcast channels.

DELAYS

t(a) = time until the next user arrives.

t(s) = service time for a user on a point-to-point channel.

t(r) = elapsed time for a re-send of connection request.

t(b) = access time for a user on broadcast channels.

As depicted by the simulation model, a mobile user first tries to connect to the base station and then requests data when he/she enters a cell. If the mobile user has valid access information for the broadcast channels, the mobile user may obtain the data of interests through the broadcast channels. Otherwise, he/she will need to obtain the data through a point-to-point channel. If the traffic is light, the mobile user can easily connect to the base station. However, if the system is overloaded, i.e., all of the available channels are in use, the mobile user will have to re-send the connection requests until one of the other mobile users relinquishes a channel.

One major advantage of this simulation model is that it accurately captures activities of the wireless information dissemination systems. By implementing the model with tools or programs, we can simulate those activities and obtain important information regarding to system characteristics and performance.

3   Multiple Cell Mobile Environments

In this section, we extend the event graph of a single cell wireless information dissemination system to cover the scenarios of channel borrowing and hand-off in multiple cell mobile environments.

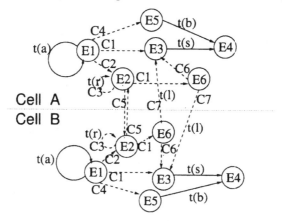

Fig. 3. Channel Borrowing

3.1   Channel Borrowing

If some cells become overloaded, it may be possible to re-allocate channels from the lightly loaded cells to the heavily loaded cells. Without losing generality, we present in Figure 3 an event graph for wireless information systems in a two cell mobile environment. The event graph for multiple cell mobile systems can be similarly derived.

EVENTS
   E1: a user attempt to connect to the base station.
   E2: blocked user re-sends connection request to the base station.
   E3: service session begins.
   E4: session ends.
   E5: user tunes into the broadcast channels.
   E6: point-to-point channel allocated.

CONDITIONS
   C1: $m < c$.
   C2: $m = c$.
   C3: restless user keeps re-sending connection requests.
   C4: user has valid access information for broadcast channels.
   C5: $(m = c$ and $m' < c')$[1] or $(m < c$ and $m' = c')$.
   C6: user from the cell.
   C7: user from other cell.

DELAYS
   t(a) = time until the next user arrives.
   t(s) = service time for a user on the point-to-point channel.
   t(r) = elapsed time for a re-send of connection request.
   t(b) = access time for a user on the broadcast channels.
   t(l) = setup time for channel borrowing.

---

[1] Assuming m' and c' are the number of mobile users and channels in the neighbor cell, respectively.

The basic activities in this multiple cell simulation model are the same as those in the one cell model. However, when all of the available channels in a cell are in use, a mobile user is allowed to obtain a point-to-point channel from the neighbor cell. The steps for channel borrowing are done through negotiation between the two base stations. Thus, the mobile user doesn't know the actions behind the scene. To capture this scenario in the extended multiple cell simulation model, transition arrows between E2 nodes of the neighbor cells are added to the event graph with a condition, C5, which states that all channels in the current cell are in use and that there are free channels in the neighbor cell. Moreover, an event E6 are separated from E3 of the single cell model to denote the situation that a channel has been allocated. The mobile user receives service from the base station of its residential cell. We use t(l) to denote the overhead for channel borrowing.

3.2 Hand-Off

Roaming is one of the most important features for mobile systems. Although the mobile users can travel from cell to cell, it is required that a service in progress not be interrupted. Therefore, change of channels and transfer of services due to change of cells, i.e., hand-off, must be conducted transparently. In the following, we again assume a two cell information dissemination system and provide an event graph for the system (see Figure 4). In this model, we assume that both cells are capable of providing the same services.

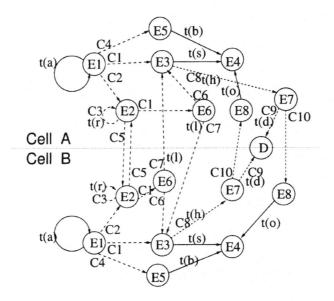

Fig. 4. Hand-off

EVENTS

E1: a user attempt to connect to the base station.
E2: blocked user re-sends a connection request to the base station.
E3: service session begins.
E4: session ends.
E5: user tunes into the broadcast channels.
E6: point-to-point channel allocated.
E7: hand-off begins.
E8: hand-off completes.
D: connection drops.

CONDITIONS

C1: $m < c$.
C2: $m = c$.
C3: restless user keeps re-sending connection requests.
C4: user has valid access information for broadcast channels.
C5: $(m = c$ and $m' < c')$ or $(m < c$ and $m' = c')$.
C6: user from the cell.
C7: user from other cell.
C8: roaming users.
C9: hand-off not successful.
C10: hand-off successful.

DELAYS

$t(a)$ = time until the next user arrives.
$t(s)$ = service time for a regular user using a point-to-point channel.
$t(r)$ = elapsed time for a re-send of connection request.
$t(b)$ = access time for a user on broadcast channels.
$t(l)$ = setup time for channel borrowing.
$t(h)$ = service time for a roaming user in departing cell.
$t(o)$ = service time for a roaming user in arriving cell.

GLOBAL CONSTRAINTS

$t(s) = t(h) + t(o)$.

The above model is a direct extension of the multiple cell model which accommodates channel borrowing. Thus, it covers both channel borrowing and hand-off activities in a cell. As described in the event graph, a user enters the event E7 when she is about to roam into the neighbor cell. From this moment, the base station negotiates with the neighbor base station to complete the hand-off process. The duration of services used in the departing cell is denoted as $t(h)$ and the duration of services used in the arriving cell is $t(o)$. Thus, the sum of above two durations is the same as the average service time by a regular user. In the model, we also use an event D to denote the situation where there is no channel available and the connection has to be dropped.

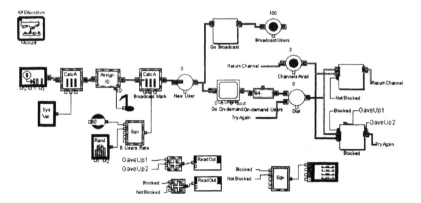

Fig. 5. Extend Model for Single Cell with Broadcast Channel

4   Simulations and Performance Evaluation

Blocking probability is one of the major performance measurements for mobile computing systems. If all of the channels are busy in a cell, a service request is blocked. When a large number of mobile users are connected to the system, the blocking probability may reach an unacceptable level. In this situation, the traffic alleviation techniques such as channel borrowing and broadcasting are necessary. In this paper, we use blocking probability to indicate the impact of burst traffic and evaluate the effectiveness of the traffic alleviation techniques.

Our experiments are conducted by using two separate implementation approaches. In order to quickly obtain simulation results, we first use an easy-to-use, advanced simulation tool, Extend [Inc95], to develop the simulation systems based on the event graph we proposed earlier. Through Extend's visual interface, we dynamically create simulation models from existing building blocks. Figure 5 shows an Extend simulation model[2] for one cell system with a broadcast channel. In addition to easy implementation, a great advantage obtained by using Extend is that the activities within the models can be observed through animations during executions. Thus, implementation errors can be easily avoided. However, the performance of our Extend-based simulation systems is rather poor. It took several hours to execute one run of our simulation on a Pentium II/450 PC, so it's not effective to conduct massive simulations for reliability and confidence tests. As a result, we implement another simulation system in Fortran/C on the Pentium PC. Finally, we use the simulation results obtained from these two separate implementations to verify their correctness.

In our experiments, system blocking probability corresponding to different distributions of mobile users arrival rate are obtained. Two distributions of mobile user arrival rate are formulated and shown in Figure 6. In the figure, the

---

[2] Due to the limited space, we only show the top level presentation of the model. There are more details in boxes such as 'Go Broadcast'.

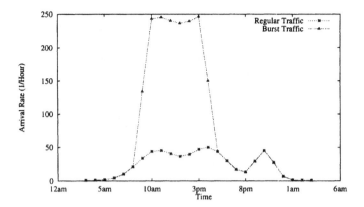

Fig. 6. Distributions of mobile user requests

curves denote the average arrival rate of mobile users during a day. *Regular traffic* represents the daily distribution of mobile users requesting for services, while *burst traffic* represents the distribution with some special events occuring, e.g., chess match between Deep Blue and Kasparov. We also compare the blocking probability based on different combinations of channel numbers and service time.

In the current state of our implementation, we have only finished the developments for single cell systems.[3] We have configured our systems based on two different channel allocations: one allocates all of the channels to point-to-point communications and the other one allocates one channel for broadcast service and leaves the rest of channels for point-to-point communications.

Monte Carlo[4] simulations are conducted on these two implementations described above. In our simulations, we assume that there are 20 channels within a cell. The users have an average service time of 20 minutes. Among blocked users, 15% of the users try to reconnect. These users have an average delay of 3 minutes for re-tries and they give up after 10 tries. For the system with broadcast channels, we assume that 15% of users are only interested in the information available on broadcast channels, so those users may tune into the broadcast channel when they enter the system. The systems simulate the activities within a cell during a period of 24 hours. We choose to have the period start at 3am, because a cell usually has the least users at the time.

Figure 7 and 8 show our simulation results on cumulative blocking probability during a day. A point at position (x, y) denotes that there are $y\%$ of mobile users blocked from 3am to x. Figure 7 shows the results collected during one run of simulation on Extend system, while Figure 8 shows the average blocking

---

[3] The extensions of simulations for multiple cells environments will be added to the next versions of our implementation.

[4] Monte Carlo methods refers the branch of experimental mathematics that deals with experiments on random numbers.

probability obtained from executing 100 runs of the monte carlo simulations on the C/Fortran system. It can be easily observed from the figures that simulations on both systems show consistent results, which makes us feel very confident about the correctness and accuracy of our implementations.

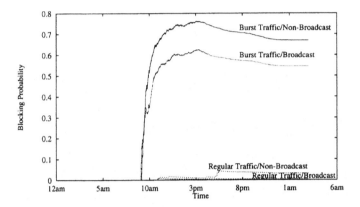

Fig. 7. Blocking probability for a day (Extend)

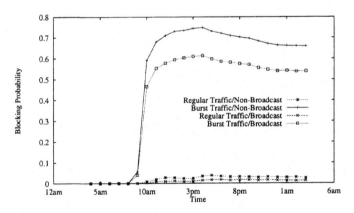

Fig. 8. Blocking probability for a day (C/Fortran)

From the figures, we can observe that the blocking probabilities for both regular and burst distributions in the first 5 hours are almost 0 since there are virtually no mobile users turned away. For the rest of the day, the blocking probability for regular distribution is kept low while the blocking probability for burst traffic jumps at around the 9am. The blocking probability for burst traffic decreases since the user arrivals are back to normal after 5pm. From the figures, it's also obvious that the blocking probabilities for mobile systems with a broadcast channel are lower than that for systems without a broadcast channel.

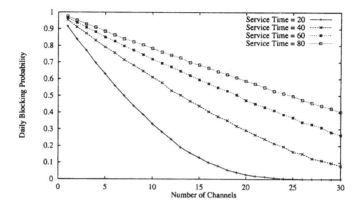

Fig. 9. Blocking probability vs number of channels/service time (non-broadcasting)

In fact, the blocking probability will be reduced even more significantly if most of the users in the burst traffic are using a broadcast channel. Thus, when a cell is experiencing overwhelming requests for connection due to some timely events, the base station should allocate channels for broadcasting information related to these events in order to serve the burst requests and relieve the load of point-to-point connections.

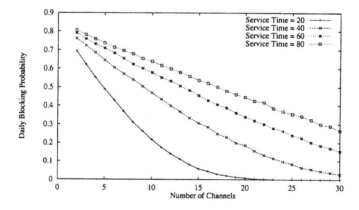

Fig. 10. Blocking probability vs number of channels/service time (broadcasting)

In order to answer questions such as "Given an average service time, how many channels are needed to provide satisfactory services (in terms of blocking probability)?", we conduct simulations to show the daily blocking probability corresponding to number of channels and service time (in minutes), for mobile systems without and with a broadcast channel, respectively. For this experiment, we ran 100 monte carlo simulations for each combination of number of channel and service time to obtain the daily blocking rate. As shown in Figure 9 and Figure 10, the blocking probability decreases as the number of channels allocated increases. Meanwhile, the blocking probability increases as the user service time increases. Once again, the figures show that the broadcast channel can reduce the blocking probablity significantly.

The above experiment results are important, because they allow us to verify the correctness of our simulations and set up a foundation for more experiments with much more complex mobile systems in the future. Also, the simulation results provide valuable data and guidance for researchers who are interested in the analytical study of the systems.

5   Conclusion

The era of ubiquitous information services is forthcoming. With increasing demands for bandwidth, information broadcasting and cell splitting techniques have been proposed to address the bandwidth and communication traffic problems. New information dissemination systems have to take broadcast channels into their designs and consider issues involved in multiple cell mobile environments, such as channel borrowing and hand-off. Performance analysis models and tools may facilitate a fundamental understanding and fine-tuning of these systems. Thus, they are critical for the business success of the systems.

In this paper, we have proposed three dynamic discrete-event based simulation models for a wireless information dissemination system, which includes both broadcast and on-demand services. Compared to previous analytical studies on similar systems, our models relax constraints assumed by queueing theory based models while providing easily understandable logic for simulations. Moreover, our models treat on-demand and broadcast services coherently and capture one of the special activities in communication systems, *retry*. Finally, and most importantly, our models address important issues in multiple cell environments, such as channel borrowing and hand-off, and thus enable multiple cell performance analysis.

Based on the dynamic discrete-event model we proposed for wireless information dissemination systems in the single cell scenario, we have implemented two simulation systems: one was quickly prototyped with a visual programming simulation tool, Extend, and the other one was programmed in C/Fortran languages. We conducted Monte Carlo simulations for accumulative blocking probability using the two systems we developed. Our experiments show that the pre-allocated point-to-point channels within a cell can easily handle the regular mobile traffic, but it may have problems handling burst traffic. In this situation, broadcast channels disseminating timely information may satisfy some of the connection demands and thus reduce the blocking rate of the system. Finally, our simulations reveal an important relationship between blocking probability and various combinations of number of channels and service time. Through the simulations, we may predict how many channels are needed to satisfy users with certain average service time. This data is critical for mobile system planners and service providers.

This work is only a start of our simulation study on wireless information systems. Through our modeling and implementations, we have set up a foundation for more experiments with much more complex systems in various mobile environments. The simulation results provide valuable information and guidance for researchers who are interested in analytical study of the systems. As the next step, we plan to extend our implementations and experiments to the multiple cell environments. The simulation implementations will serve as our test bed to address issues in designing information services for multiple cell environments, such as global broadcast scheduling, service hand-off, multiple cell cache optimization and mobile proxy management.

References

[AAFZ95] S. Acharya, R. Alonso, M. Franklin, and S. Zdonik. Broadcast disks: Data management for asymmetric communications environments. In *Proceedings of the ACM SIGMOD Conference on Management of Data*, San Jose, California, 1995.

[BB95] A. Bakre and B. R. Badrinath. Handoff and systems support for indirect tcp/ip. In *2nd Usenix Symposium on Mobile and Location-Independent Computing*, April 1995.

[HP89] S.V. Hoover and R.F. Perry. *Simulation: a problem-solving approach.* Addison-Wesley, New York, 1989.

[IB94] T. Imielinski and B. R. Badrinath. Wireless mobile computing : Challenges in data management. *Communication of ACM*, 37(10), 1994.

[Inc95] Imagine That Inc. *Extend – Performance Modeling for Decision Support.* Imagine That, Inc., CA, 1995.

[IV94] T. Imielinski and S. Viswanathan. Adaptive wireless information systems. In *proceedings of SIGDBS (Special Interest Group in DataBase Systems) Conference*, Tokyo - Japan, October 1994.

[LHL99] W.-C. Lee, Q.L. Hu, and D. L. Lee. A study of channel allocation methods for data dissemination in mobile computing environments. *ACM/Baltzer Mobile Networks and Aplications (MONET): Special Issue on Resource Management in Wireless Networks*, 4(2):117–129, 1999.

[SRB97] K. Stathatos, N. Roussopoulos, and J. S. Baras. Adaptive data broadcast in hybrid networks. In *Proceedings of the 23rd VLDB Conference*, pages 326–335, Athens, Greece, 1997.

[TJ91] S. Tekinay and B. Jabbari. Handover and channel assignment in mobile cellular networks. *IEEE Communications Magazine*, pages 42–46, November 1991.

Session II: Transaction Processing in Mobile Environments

On the Performance of Transaction Processing in Broadcast Environments

Victor C. S. Lee[1], Sang H. Son[2], and Kwok-Wa Lam[1]

[1] Department of Computer Science, City University of Hong Kong
csvlee@cityu.edu.hk
[2] Department of Computer Science, University of Virginia
son@cs.virginia.edu

Abstract. In many mobile computing systems, most of the transactions are read-only. In this paper, we propose an algorithm to process read-only transactions in broadcast environments such that the serializability of transactions is maintained. The serializability of transactions is a crucial issue in applications such as stock trading. However, in broadcast environments, the upstream communication capacity from mobile clients to the server is very limited. Therefore, conventional concurrency control protocols, which require equal bandwidth of communication on both sides of mobile clients and the server, become handicapped in this environment. In our algorithm, read-only transactions can be completed locally and autonomously at the mobile clients without upstream communication, which is a highly desirable feature for the scalability of applications running in broadcast environments. The simulation results show that our proposed algorithm performs well in a wide range of settings.

1 Introduction

In the near future, tens of millions of users will be carrying a portable computer that uses a wireless interface to access a world-wide information network for business or personal use [6]. Broadcast-based data (such as stock price) dissemination is likely to be a major mode of information transfer in wireless environments [2], [6], [13]. Some important applications in broadcasting environments are information dispersal systems and information retrieval systems where read-only transactions [5] are dominant. For example, in stock trading applications, it is expected that millions of brokers might read the prices of multiple stocks for the computation of composite index before they decide to buy any stock.

Broadcast disks [1] are a form of data dissemination systems that are well suited for wireless and mobile computing environments. The server continuously and repeatedly broadcasts all data objects in the database. The mobile clients view this broadcast as a disk and can read the values of data objects being broadcast. A periodic broadcast program is constructed to schedule the broadcast of data objects cyclically according to certain popularity criteria. Some unused extra broadcast slots in each broadcast cycle can be used to broadcast additional information such as the control information described below in our algorithm.

H.V. Leong et al. (Eds.), MDA'99, LNCS 1748, pp. 61-70, 1999.
© Springer-Verlag Berlin Heidelberg 1999

Existing concurrency control protocols for transaction processing are not suitable for broadcast environments. One of the reasons is the requirement of bi-directional communications between the clients and the server such as in most client-server systems. Another reason is a large number of clients in broadcast environments. The server may be overloaded by simultaneous submission of large number of lock requests. The blocking resolution used by the locking-based protocols may delay the execution of the blocking transactions. This is even worse in broadcast environments where the upstream bandwidth from the clients to the server is limited.

Data management in broadcast environments receives a lot of attention in these few years [3], [4], [6], [8], [11]. However, there are only a few studies on transaction processing. In [11], a correctness criterion is proposed to allow read-only transactions to read current and consistent data in broadcast environments without contacting the server. However, the serializability is not maintained in their protocol. Two different read-only transactions may perceive the effects of update transactions in different serialization orders. It may be hazardous to certain applications such as mobile stock trading where a buy/sell trade will be triggered to exploit the temporary pricing relationships among stocks. In [12], they proposed two protocols, F-Matrix and R-Matrix. Although the F-Matrix shows better performance, it suffers from high overheads in terms of expensive computation and high bandwidth requirement for additional control information for consistency check.

In [9], [10], a number of broadcast methods is introduced to guarantee correctness of read- only transactions. The multiversion broadcast approach broadcasts a number of versions for each data item along with the version numbers. This method increases considerably the size of the broadcast cycle and accordingly response time. Moreover, the serialization order is fixed at the beginning of the read-only transaction. It is too restrictive and lacks flexibility. For the conflict-serializability method, both the mobile clients and the server have to maintain a copy of the serialization graph for conflict checking. It incurs high overheads to maintain the serialization graph. The integration of updates into the local copy of the serialization graph and the cycle detection may be too computation intensive for certain portable mobile computers.

2   A Case for Serializability

Due to the asymmetric communication bandwidth between the mobile clients and the server, existing concurrency control protocols for transaction processing are not suitable for broadcast environments. Hence, recent work [11], [12] relaxed the strictness of serializability and proposed some concurrency control protocols based on the relaxed consistency requirements for broadcast environments. While these protocols are useful in some applications, serializability may still be needed to guarantee the correctness of some applications in broadcast environments.

To illustrate the importance of the serializability in transaction processing in broadcast environments, let's take mobile stock trading as an example. Consider two stock-trading read-only transactions Q_1 and Q_3 at mobile clients that read the stock prices of X and Y and compute composite indices of stocks X and Y with different weightings. Let U_2 and U_4 be update transactions at the server that update the prices of X and Y respectively. Assume that the price of stocks X and Y is both 100. Consider the following execution schedule:

$$r_1(X)w_2(X)c_2r_3(X)r_3(Y)w_4(Y)c_4r_1(Y)\ldots.$$

If both Q_1 and Q_3 commit, then the server and both mobile clients would *see* serializable executions. The partial serialization order at the server is $U_2 \rightarrow U_4$ whereas the serialization orders for Q_1 and Q_3 at the mobile clients are $U_4 \rightarrow Q_1 \rightarrow U_2$ and $U_2 \rightarrow Q_3 \rightarrow U_4$ respectively. Assume that the execution schedule is allowed and the stock price for X rises to 110 and is updated by U_2 whereas the stock price for Y drops to 90 and is updated by U_4. Since Q_1 and Q_3 perceive different serialization orders, Q_1 will get a composite index assuming that stock Y drops and stock X remains unchanged whereas Q_3 will get a composite index assuming that stock X rises and stock Y remains unchanged. As a result, the decision outcome of Q_1 may be to shift the cash from Y to X and that of Q_3 may be to shift the cash from X to Y, which is confusing. In fact, the global execution schedule $(Q_1 \rightarrow U_2 \rightarrow Q_3 \rightarrow U_4 \rightarrow Q_1)$ is not serializable and should not be allowed.

3   The BCC-TI Algorithm

In this section, we present the Broadcast Concurrency Control using Timestamp Interval (BCC-TI) algorithm for read-only transactions. Note that concurrency control of update transactions is maintained by the underlying conventional concurrency control protocol.

3.1   Commit Phase of Update Transactions at the Server

For every update transaction, a final timestamp is assigned when it commits. Let $WS(U)$ be the write set of an update transaction, U, and $TS(U)$ denotes the final timestamp of U. Also, let $WTS(d)$ be the largest timestamp of committed update transaction that has written data object, d. Then, when U commits, the following steps will be performed. The control information table is used for the serializability check by read-only transactions (to be described later).

Step 1: assign the current timestamp to $TS(U)$.
Step 2: copy $TS(U)$ into $WTS(d)$, $\forall d \in WS(U)$.
Step 3: store $TS(U)$ and $WS(U)$ into the control information table.

3.2   Read Phase of Read-Only Transactions at the Mobile Clients

Every read-only transaction in the read phase is assigned a timestamp interval, which is used to record temporary serialization order induced during the execution of the transaction. At the start of execution, the timestamp interval of a read-only transaction is initialized as $[0, \infty)$, i.e., the entire range of timestamp space. Whenever a serialization order of a read-only transaction is induced by its read operations or the serializability check against update transactions, its timestamp interval is adjusted to reflect the dependencies among the transactions.

The detection of non-serializable execution by read-only transactions is achieved by using the timestamp intervals. Since the serialization order of read-only transactions is checked against committed update transactions, we make use of the timestamps of data objects, i.e. $WTS(d)$ of data object d so that read-only transactions can serialize with committed transactions locally. In the read phase, whenever a read-only transaction reads a data object, its timestamp interval is adjusted to reflect the serialization induced between itself and committed update transactions. Let $LB(Q)$ and $UB(Q)$ denotes the lower bound and the upper bound of the timestamp interval of a read-only transaction, Q, respectively. If the timestamp interval shuts out $(LB(Q) \geq UB(Q))$, a non-serializable execution performed by the read-only transaction is detected, and the transaction restarts. The following procedure is performed whenever a read-only transaction, Q, reads a data object, d.

Step 1: read (d);
Step 2: $LB(Q) = max(LB(Q), WTS(d))$;
Step 3: if $LB(Q) \geq UB(Q)$ then restart (Q).

3.3   Data Broadcast

During each broadcast cycle, the server broadcasts the control information along with the data objects that programmed for the current cycle. The control information consists of the timestamps and the write sets of committed update transactions during the last broadcast cycle. The control information during each cycle helps mobile clients to determine whether read-only transactions have introduced non-serializable execution with respect to the committed update transactions. There is only one possible type of conflict between the committed update transactions and the active read-only transactions: write- read conflict (e.g., $WS(U) \cap RS(Q) \neq \{\}$). In this case, the serialization order should be $Q \rightarrow U$. That is, the timestamp interval of Q is adjusted to precede that of U. It implies that the writes of U have not affected the read phase of Q.

Before read operations are performed on data objects that are broadcast during a cycle, all read- only transactions have to consult the control information received during that cycle to determine whether the execution can proceed. If the execution cannot proceed, the read-only transaction is aborted. For each read-only transaction, Q, at the mobile client, the following step is followed before the remaining read operations. In the step, the adjustment of timestamp intervals of active read-only transactions is performed. Let CUT be the set of update transactions that are committed in the last broadcast cycle. Let $CRS(Q)$ be the current read set of Q. In other words, $CRS(Q)$ is the set of data objects that have been read by Q from previous broadcast cycles. Note that $TS(U)$ and $WS(U)$ are stored in the control information broadcast by the server.

Step 1: $UB(Q) = min_i(TS(U_i), UB(Q))$,

$\forall U_i \in CUT$ and $WS(U_i) \cap CRS(Q) \neq \{\}$;

Non-serializable execution is detected when the timestamp interval of an active read-only transaction shuts out. The non-serializable execution is aborted by restarting the read- only transaction.

4   Effectiveness of the New Algorithm

The algorithm offers autonomy between the mobile clients and the server. At mobile clients, read-only transactions can read current and consistent data object off the air without contacting the server, while the serializability is maintained. Once all the required data objects are read by a read-only transaction, the transaction can commit without communicating with the server.

Another useful feature is the flexible adjustment of serialization order of a read-only transaction with respect to update transactions. That is, the serialization order of transactions is independent of the arrival order of transactions in the validation phase, which is an implicit assumption adopted by most validation schemes. This assumption leads to unnecessary restarts [7]. By using the flexible adjustment of serialization order, unnecessary restarts can be avoided.

However, these nice features do not come for free. The main cost for those benefits is the management of timestamp intervals for active read-only transactions and the cost for the maintenance and broadcast of the control information table. The management of timestamp intervals for active read-only transactions can be done efficiently by using a transaction table at the mobile clients. The transaction table contains information of active read-only transactions, including the current read set and the upper and lower bounds of the timestamp interval of every active read-only transaction. The information that is recorded in the control information table includes the timestamp, $TS(U)$, and the write set, $WS(U)$, of each committed update transaction U in the last broadcast cycle. However, the amount of control information is much less than that of other protocols in the literature [9], [12].

5   Performance Evaluation

The simulation experiments are aimed at studying the performance of our proposed algorithm (BCC-TI) and the F-Matrix algorithm proposed in [12] for broadcast disk environments. Note that the F-Matrix maintains the update consistency (weaker than serializability) while our algorithm maintains the serializability. We also expect to identify in what range of settings one algorithm performs better than the other. We have not considered the effects of caching in this performance study. In other words, a read-only transaction may have to wait for the requested data object in the next broadcast cycle if the data object is missed (have been broadcast) in the current broadcast cycle.

The performance of these algorithms can be compared by the transaction response time, which includes the time involved in restarting the transaction. Note that a transaction may be restarted more than once. The overhead to restart a read-only transaction at a mobile client is low in both algorithms. Since the processing of the read-only transaction at the mobile clients is completely transparent and independent to the server, no additional overhead will be generated at the server while read-only transactions are restarted at the mobile clients.

5.1 Experimental Setup

To have a fair comparison with F-Matrix, our simulation model is the same as in [12]. The simulation model consists of a server, a client, and a broadcast disk for transmitting both the data objects and the required control information. The mobile clients only process read-only transactions while the server processes update transactions as well. The data objects that the transactions access are uniformly distributed in the database. Table 1 lists the baseline setting for the simulation experiments. The time unit is in bit-time, the time to transmit a single bit. For a broadcast bandwidth of 64 Kbps, 1 M bit-times is equivalent to approximately 15 seconds. The server fills the broadcast disk with the data

Table 1. Baseline Setting

Parameter	Value
Mobile Clients	
Transaction Length	
(Number of read operations)	4
Mean Inter-Operation Delay	65,536 bit times (exponentially distributed)
Mean Inter-Transaction Delay	131,072 bit times (exponentially distributed)
Concurrency Control Protocol	BCC-TI
Server	
Transaction Length	
(number of operations)	8
Transaction Arrival Rate	1 per 250,000 bit-times
Read Operation Probability	0.5
Number of Data Objects in Database	300
Size of Data Objects	8,000 bits
Timestamp Size	8 bits
Concurrency Control Protocol	Optimistic Concurrency Control with Forward Validation (OCC-FV)

at the beginning of a cycle. Each cycle consists of a broadcast of all the data objects in the database along with the associated control information. For F-Matrix, the control information consists of an $n \times n$ matrix, where n is the number of data objects in the database. For instance, the size of the matrix is 720 kbits in the baseline setting. For BCC-TI, the control information consists of a table containing the timestamps and write sets of the committed update transactions during the last broadcast cycle and an array of length n containing the write timestamps of the data objects. Each element in the array is broadcast along with the corresponding data object.

The response times are measured in bit-time and 90% confidence intervals were obtained with widths less than 5% of the point estimates of the response times.

5.2   Effect of Database Size

There are two main components in the response time of read-only transactions in mobile clients. The first one is the waiting time for the data objects. The waiting time increases with the length of the broadcast cycle. The second one is attributed to transaction restarts. Transaction restarts are mainly due to data conflicts arisen during a broadcast cycle.

For the effect of the database size, there are two counter factors affecting the response time of read-only transactions at mobile clients. When the database size increases, the length of the broadcast cycle increases, which in turn increases the waiting time for the data objects. The increase in the control information to be sent also increases the length of the broadcast cycle. Moreover, when the length of the broadcast cycle increases, the number of update transactions committed at the server per broadcast cycle increases. This increases the number of data conflicts, which leads to higher transaction restart rates. On the contrary, as the number of data objects in the database increases, the probability of the transactions accessing the same data object decreases. Subsequently, the probability of data conflicts between transactions is reduced and it leads to lower transaction restart rates.

In Fig. 1, we find that the response times increase with the database size. Clearly, the increase of the waiting time and the data conflicts due to the increase of the length of broadcast cycle become the dominating factors. Fig. 1 also shows that the BCC-TI algorithm outperforms the F-Matrix algorithm for the whole range of database size in the experiment. The amount of improvement increases with the database size. When there are 400 or more data objects in the database, the response time improvement is more than 25%. The rate of increase in response time for F-Matrix is also faster than BCC-TI. This is due to the overheads of the large control information, which leads to a longer broadcast cycle.

5.3   Effect of Data Object Size

The performance of the algorithms in terms of response time with respect to the data object size is shown in Fig. 2. The length of the broadcast cycle increases with data object size and hence the response time increases. BCC-TI still outperforms F-Matrix for a wide range of the data object size. As the data object size increases, the relative overhead due to control information tends to decrease for F-Matrix. Hence, the two algorithms approach each other as data object size increases.

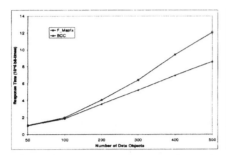 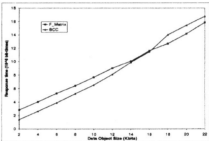

Fig. 1. Response Time vs. Database Size   **Fig. 2.** Response Time vs. Object Size

On the other hand, the increase in the length of broadcast cycle increases the number of data conflicts. However, different from the case of database size, the data object size does not affect the access frequency of data objects, which in turn affects the number of data conflicts. Since BCC-TI maintains the serializability while F-Matrix maintains the update consistency, BCC-TI is more vulnerable to the increase in the number of data conflicts. When the length of broadcast cycle is long, it is very likely for a data object that has been read by a read-only transaction at a mobile client be written by an update transaction at the server.

In spite of a relatively higher probability of data conflicts, BCC-TI outperforms F-Matrix in terms of response time until the size of a single data object is huge (>2,000 bytes). In other words, the saving of the overheads of transmitting the $n \times n$ matrix, where n is the number of data objects in the database, outweighs the loss of time in restarting read-only transactions at mobile clients due to data conflicts. Moreover, the overheads of computing the matrix are also removed. In fact, the bandwidth resource in broadcast environments is more worth saving than the transaction restart overheads at mobile clients.

5.4   Effect of Server Transaction Arrival Rate

Another factor that will increase data conflict is transaction arrival rate at the server. The higher the server transaction arrival rate, the more likely there are data conflicts. Although there may be data conflicts among transactions at the server, we focus on the data conflicts with the mobile transactions. Fig. 3 shows the response time of the mobile transactions with respect to the server transaction arrival rate. It is obvious that the response time increases with the arrival rate but the magnitude is far smaller than the impact of database size or data object size. It is because the server transaction arrival rate will not affect the length of broadcast cycle. The sole effect of data conflicts leading to mobile transaction restarts is less than the combined effect of increased length of broadcast cycle leading to increased waiting time and mobile transaction restarts.

To maintain the serializability, BCC-TI is more sensitive to data conflicts than F-Matrix. When the server transaction arrival rate is high, BCC-TI begins to lose ground to F-Matrix.

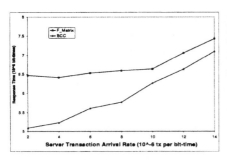

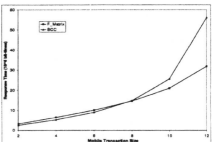

Fig. 3. Response Time vs. Server Transaction Arrival Rate

Fig. 4. Response Time vs. Mobile Transaction Size

5.5 Effect of Mobile Transaction Size

The transaction size at mobile clients is usually not very large. It is expected that long mobile transactions will be relatively slow and vulnerable to data conflicts with transactions at the server. Nevertheless, we would like to compare the performance of both algorithms by experimenting with a wide range of mobile transaction size. In Fig. 4, the response time of both algorithms rises as the mobile transaction size increases. When the mobile transaction size is smaller than 8 operations, the performance of BCC-TI is better than F-Matrix by about 20%. However, the performance of BCC-TI loses when the mobile transaction is long. Note that there is no cache in the simulation model, it is very likely for a mobile transaction to read different data objects in different broadcast cycles and a long mobile transaction may need to read through a number of broadcast cycles. As a result, the probability of data conflict increases and it is even worse for BCC-TI that maintains the serializability. Note that when the size of mobile transactions is greater than 8, the response time of both algorithms may be outside an acceptable range (>20M bit-times or 300 seconds).

6 Conclusions and Future Work

In this paper, we propose an algorithm, broadcast concurrency control with timestamp interval (BCC-TI), to maintain the serializability of read-only transactions at mobile clients in broadcast environments. BCC-TI offers a number of highly desirable features for applications running in broadcast environments. First of all, the serializability of read-only transactions can be maintained. Read-only transactions can be committed locally at mobile clients without upstream communication. BCC-TI can be easily integrated with other conventional concurrency control protocols that are employed to maintain the serializability at the server. Moreover, the flexible adjustment of serialization order of a read-only transaction with respect to update transactions helps to reduce the number of unnecessary restarts.

We are now studying how to process update transactions, in addition to read-only transactions, at mobile clients effectively. We are also interested in the issue of cache management in broadcast environments.

References

1. Acharya, S., Alonso, R., Franklin, M., and Zdonik, S., ŰBroadcast Disks: Data Management for Asymmetric Communication Environments,Ţ Proceedings of the ACM SIGMOD Conference, pp. 199-210, 1995.
2. Alonso, R., and Korth, H., Database Systems Issues in Nomadic Computing, Proceedings of the ACM SIGMOD Conference, Washington D.C., pp. 388-392, 1993.
3. Barbara, D., and Imielinski, T., Sleepers and Workaholics: Caching Strategies in Mobile Environments, Proceedings of the 1994 ACM SIGMOD International Conference on Management of Data, pp. 1-12, 1994.
4. Dunham, M. H., Helal, A., Balakrishnan, S., ŰA mobile transaction model that captures both the data and movement behavior,Ţ Mobile Networks and Applications, vol. 2, pp. 149 ô 162, 1997.
5. Garcia-Molina, H., and Wiederhold, G., Read-Only Transactions in a Distributed Database, ACM Transactions on Database Systems, vol. 7, no. 2, pp. 209-234, 1982.
6. Imielinski, T., and Badrinath, B. R., ŰMobile Wireless Computing: Challenges in Data Management,Ţ Communications of the ACM, vol. 37, no. 10, pp. 18-28, 1994.
7. Lee, J., and Son, S. H., Using Dynamic Adjustment of Serialization Order for Real-Time Database Systems, Proceedings of 14th IEEE Real-Time Systems Symposium, pp. 66-75, 1993.
8. Pitoura, E., and Bhargava, B., ŰBuilding Information Systems for Mobile Environments,Ţ Proceedings of the third international conference on Information and knowledge management, pp. 371-378, 1994.
9. Pitoura, E., ŰSupporting Read-Only Transactions in Wireless Broadcasting,Ţ Proceedings of the DEXA98 International Workshop on Mobility in Databases and Distributed Systems, pp. 428-433, 1998.
10. Pitoura, E., and Chrysanthis, P. K., ŰScalable Processing of Read-Only Transactions in Broadcast Push,Ţ Proceedings of the 19th IEEE International Conference on Distributed Computing System, 1999.
11. Shanmugasundaram, J., Nithrakashyap, A., Padhye, J., Sivasankaran, R., Xiong, M., and Ramamritham, K., ŰTransaction Processing in Broadcast Disk Environments,Ţ Advanced Transaction Models and Architectures, Jajodia, S., and Kerschberg, L., editors, Kluwer, Boston, pp. 321-338, 1997.
12. Shanmugasundaram, J., Nithrakashyap, A., Sivasankaran, R., and Ramamritham, K., ŰEfficient Concurrency Control for Broadcast Environments,Ţ ACM SIGMOD International Conference on Management of Data, 1999.
13. Zdonik, S., Alonso, R., Franklin, M., and Acharya, S., ŰAre Disks in the Air Just Pie in the Sky,Ţ Proceedings of the Workshop of Mobile Computing Systems and Applications, California, 1994.

Transaction Processing in an Asymmetric Mobile Environment*

Eppie Mok[1], Hong Va Leong[1], and Antonio Si[2]

[1] Department of Computing, Hong Kong Polytechnic University, Hong Kong.
{cseppie,cshleong}@comp.polyu.edu.hk
[2] Sun Microsystems, 901 San Antonio Road, Palo Alto, CA 94303.
asi@eng.sun.com

Abstract. Transaction processing is useful in a mobile environment. The inherent support for broadcast channels together with conventional point-to-point paradigm dictates different mode of transaction processing in the mobile context. To improve system scalability, a transaction should utilize relatively high bandwidth broadcast channels for its read operations on hot data items. Read operations on cold data items and update operations would be carried out via point-to-point channels on demand. We devise a concurrency control protocol that exploits both types of channels. We address the disconnection problem by caching data items needed at a mobile client, performing transactions locally, and committing them upon reconnection. The performance of our protocol is compared with other protocols through simulation.

1 Introduction

Advances in wireless communication technology have fostered mobile computing and mobile data access [9]. We consider a mobile data access environment as a collection of base stations or servers connected via fixed wired network and a number of mobile clients accessing a database server via wireless network in the form of transactions.

There are two communication paradigms in a mobile environment: broadcast paradigm and point-to-point [2] paradigm. To improve system scalability, broadcast paradigm is adopted to disseminate data items of common interest to mobile clients, due to the high cost of having a mobile client to contact the server for data items. Various approaches have been proposed to improve data access efficiency over a broadcast channel [1, 13].

Although a broadcast channel has a higher bandwidth, it is an asymmetric one-way communication media; thus, point-to-point channels are needed to support updates to database [2]. We envision a system utilizing both paradigms to be effective and scalable, by scheduling data items of high affinity over broadcast channel while allowing update and access to items of low affinity via point-to-point channels. Performance metrics include *tuning time* and *access time* [10].

---

* This research is supported in part by the Hong Kong Polytechnic University research grant number 351/496.

H.V. Leong et al. (Eds.), MDA'99, LNCS 1748, pp. 71–81, 1999.

Listening to the broadcast channel requires a client to be in an active mode and thus, increases power consumption. Selecting popular data items to be broadcast could minimize the access and tuning time.

A mobile environment is weakly-connected; mobile clients may disconnect themselves to conserve energy. To improve operability in disconnected mode, a mobile client has to cache frequently accessed data items in its local storage, used upon disconnection. Since the cache size is limited, cache management algorithms should be considered. This has been studied in file system [11] and mobile database access [5]. The effectiveness of a caching scheme during disconnection can be characterized by *cache hit ratio* and *self-answerability* [12].

In the presence of updates initiated by mobile clients, transactional consistency to data items is required. Based on the broadcast nature of a mobile environment, researchers considered the processing of read-only transactions, making use of the consistency of items within a broadcast cycle, due to asynchronous updates at database server. In [14], multi-versioning of broadcast data items is adopted to improve the commit rate of read-only mobile transactions. Update mobile transactions are considered in [15]. With updates from mobile clients, a back channel is used to convey them back to the server, where the updates are validated and installed. In [3], a kind of certification for the updates is performed at the server, a variant of optimistic concurrency control [4]. Recognizing the constraint of serializability as the correctness criteria, it is relaxed to update consistency and the *APPROX* algorithm, with various implementations, is proposed [17].

In this paper, we propose to utilize not only the broadcast channel to disseminate data items, since a broadcast cycle can be long for a typical database. Rather, we assume a typical 80/20 data distribution and broadcast only the 20% "hot" data items, leaving the remaining 80% "cold" items to be read by clients on demand. Updates are propagated back to server for validation and installation. To keep a broadcast cycle short, we phase in a consistent data version for every broadcast. At most two data versions are maintained at the server and the broadcast cycle is made up of only one version, rather than multiple versions as in [14]. We enforce serializability instead of a weaker notion as in [17]. To cater for disconnected operation, mobile clients will cache data items, with versioning information. Read-only transactions can be processed during disconnection, contributing to self-answerability. Transactions are "pseudo-committed" during a disconnection and reintegrated with the database upon reconnection.

This paper is organized as follows. In Sect. 2, we outline our data access model and concurrency control protocol. In Sect. 3, we describe our simulation model for performance study, with evaluation results described in Sect. 4. This paper is concluded with discussion on our future research work in Sect. 5.

2   Design of Protocol

We exploit the asymmetric communication behavior of a mobile environment and focus on two different research aspects. First, hot data items at the server need to be organized over a broadcast channel appropriately. This has a direct impact on the access procedure a mobile client uses to read data items. Second, an interaction protocol between mobile clients and the server is needed for update operations as well as read operations on cold data items. This protocol defines how transactions are to be processed and the conditions for them to be committed/aborted. We will also consider the actions at a client when it becomes disconnected as well as upon reconnection.

For notational convenience, a mobile client is denoted by S_i, defined by its unique IP or Ethernet address. The set of transactions initiated by S_i is denoted by $\{T_{i1}, T_{i2}, ..., T_{im}\}$, where the local sequence number of transaction can be generated by a counter. Each data item, x, stored at the server is associated with an ID, and a version number, which is the commit timestamp, t, of the transaction, T_{ij}, that updates x. This version is denoted as x_t. To facilitate identifying the transaction that updates x, the versioning information for x_t also contains the transaction ID, i.e., i and j. Finally, each cached copy of x at a client is associated with an additional field: the broadcast cycle number, c, when x is cached. This is useful when a client operates in disconnected mode.

In our approach, each mobile client implements both memory and storage caching. When a client receives or updates a data item, it is first cached in local memory. The LRU algorithm is employed to replace data items when the memory cache is exhausted. Replaced dirty items will be written back to the storage cache, which is also managed using the LRU algorithm. During disconnection, both memory cache and storage cache contribute to the availability of data items.

2.1   Broadcast Organization and Access

The set of data items to be broadcast is selected based on popularity. We assume the typical 80/20 rule for data item affinity, such that the top 20% data items are included in a broadcast cycle. For simplicity, we assume that the hot spot is static. Adaptation to changes in hot spot for broadcast databases has been considered in [13]. We employ an indexing structure to organize the items over a broadcast channel. Several data items are organized into a bucket, with global indexing information being broadcast at the beginning of the bucket. The bucket header may further contain information about each data item. We adopt the $(1,m)$ indexing scheme [10] to organize the broadcast items, and the bucket is the unit to be broadcast.

To reduce the size of a broadcast cycle, we broadcast only one version of data items in each cycle. Furthermore, we make sure that each cycle represents a *consistent snapshot* of the database. This means that if there were an imaginary read-only transaction that reads all broadcast items, the values returned by the transaction would be consistent, since all concurrent update transactions are either serialized before that read transaction or after it. Thus, those values could be broadcast within the same cycle, and we call them the "current" version. A technique similar to shadow paging can be applied to manage the data versions. Successful updates are normally made to "shadow" version, which is not being broadcast. However, if *all* updates of a transaction to the broadcast items are made solely to those items not yet transmitted over the broadcast channel at the moment, the updated values would be installed as the "current" version, thereby improving data currency without violating consistency of items within a broadcast cycle. At the end of a broadcast cycle, those updated shadow versions will become the current versions for the new broadcast.

A connected mobile client will read its data items through the broadcast channel if they belong to the broadcast list. As an optimization, several read requests originating from the same client could be batched together before tuning to the broadcast channel. This allows the items to be retrieved in the order they are broadcast, rather than in the order the requests are issued. This not only reduces the tuning time by accessing the index only once, but also greatly reduces the access time by avoiding listening to multiple broadcast cycles.

When the required data item appears in the broadcast cycle, the client may choose to download the item immediately, or check if the version is a newer one before downloading it. For small data items, downloading without checking is simple, but for large data items, checking for their version numbers before downloading pays off, by saving energy from tuning for the unnecessary duplicated items. We will adopt the checking strategy for better performance.

2.2   Point-to-Point Channel Data Request

A connected mobile client can request a "cold" data item not belonging to the broadcast list by sending a message containing information about the transaction, T, to the server. The server will select the most recent consistent version of the requested cold items, with respect to potential concurrent updates made to items that T has read and that T is going to read, and return to the client. The server keeps track of the versions of data items that T reads. Whenever it receives a data request, it will first check if the most recent version violates consistency from the client's perspective. If the read is allowed, it sends the most recent version to the client and adds this information to a *read set*. If the current version is not consistent, the older version is sent instead. The consistency check at the server is same as *Phase 1* of the validation of a transaction, as shown in Fig. 1. If both copies of a requested data item in the server are not consistent, a client will try to read the item from its local cache. If the item is not in the cache, the transaction will be aborted. Thus, items read over point-to-point channels are consistent, so are those read over the broadcast channels. Thus, a client only needs to verify that both types of items are consistent with each other before committing the transaction.

Upon receiving the return values for its read request, a mobile client may have a choice to download the values immediately or download the values only if the items are fresher than their cached values at the client. As with broadcast, it is advantageous to perform the test for large items before downloading. We thus, adopt the checking strategy for better performance.

It can be observed that sending a data item via a point-to-point channel is not required, if the same version of the item is already cached in the client. One could augment the request message with the version number of every cached data item the client requests, so that the server can avoid sending a duplicated item, thus saving the scarce bandwidth. The server sends the result followed by the valid version after the consistency check. The client reads the item and caches it if that is a valid copy. Although there is a small overhead by including the version number in the return message, the saving due to avoiding duplicates can be substantial.

2.3   Validation and Write Phase

We adopt the deferred update approach by buffering all updates at the client. The updated values will be sent to the server for validation at the end of a transaction and will only be committed if the validation succeeds. We propose a protocol to validate transactions in mobile database. The idea is similar to optimistic concurrency control with dynamic adjustment of serialization order [4], though that protocol is not designed for multi-versioned data and mobile environment, where clients may disconnect and reconnect frequently. We generalize it to utilize several data versions, with Thomas's Write Rule and apply concept from Coda file system [11] to support disconnected operations. Transactions can have their committed timestamp adjusted as long as conflicting transactions can be serialized. In our multi-version optimistic concurrency control protocol with timestamp adjustment, maintaining two data versions at server improves the chance of reading a correct value, in addition to the benefit brought about by consistency among items within each broadcast cycle. It ensures atomic property and one-copy serializability.

Upon termination of a transaction T, a mobile client submits a commit request with client and transaction ID, read set and write set to the server. As an optimization, the read set for items retrieved over the point-to-point channel can be omitted, since the server knows about them. When the server receives the commit request for T, it performs validation as detailed in Fig. 1. Upon successful validation, the updates are applied to the database, either inside a critical section or using a conservative two phase locking protocol. Data will only be written to the database if the maximum timestamp of the transactions that write x is less than the timestamp of the submitted transaction.

2.4   Disconnected Operation

When a mobile client is disconnected, read operations could still be serviced if the requested data items exist in the cache of the client. If the items do not exist in the cache, the transaction is aborted. Upon a write operation, the value is written to the temporary copy and local copy is overwritten when the transaction finishes. If all data items read by a read-only transaction, T, come from the same previous broadcast cycle or are previously written by a same transaction, T can be committed, since the data items read are consistent. However, T will not be assigned any timestamp. For update transaction T', a local validation is performed. T' is aborted if it fails the validation. Otherwise, it enters the pseudo-commit state. In other words, T' can be considered committed, but whether it

Data structures:
    tnc: largest transaction ID of successful transaction; used for timestamping.
    T.RS, T.WS: set of data items that T has read/written.
    T.TS: commit timestamp of T.
    lts, lts$'$, uts, uts$'$: lower/upper bound of possible commit timestamp.

To commit transaction T:
Phase 1: check for consistent read set
lts $\leftarrow$ 0; uts $\leftarrow$ ∞
foreach data item $x \in T$.RS do
    let T_a be the transaction that creates x
    lts $\leftarrow$ max(lts, T_a.TS)
endfor
foreach $x \in T$.RS do
    let T_a be the transaction that creates x
    for t from T_a.TS + 1 to lts do
        let T_b be the transaction committed with timestamp t
        if $x \in T_b$.WS then valid $\leftarrow$ false
    endfor
endfor

Phase 2: check for read write conflict (backward adjustment)
foreach $x \in T$.RS do
    for t from lts + 1 to tnc do
        let T_a be the transaction committed with timestamp t
        if $x \in T_a$.WS then
            if $t <$ uts then
                uts $\leftarrow$ t // timestamp adjustment
                if uts $\leq$ lts then valid $\leftarrow$ false
            endif
        endif
    endfor
endfor

Phase 3: check for write read conflict (forward adjustment)
lts$'$ $\leftarrow$ lts; uts$'$ $\leftarrow$ uts
foreach $x \in T$.WS do
    for t from lts + 1 to tnc do
        let T_a be the transaction committed with timestamp t
        if $x \in T_a$.RS then
            let T_b be the transaction with timestamp t' creating the version $x_{t'}$ that T_a reads
            lts$'$ $\leftarrow$ max(lts$'$, t); uts$'$ $\leftarrow$ min(uts$'$, t')
        endif
    endfor
endfor
if uts$'$ $>$ lts then
    if lts$'$ $<$ uts then lts $\leftarrow$ lts$'$ else uts $\leftarrow$ uts$'$
else
    if lts$'$ $<$ uts then lts $\leftarrow$ lts$'$ else valid $\leftarrow$ false
endif

Phase 4: assign timestamp
if valid then
    tnc $\leftarrow$ tnc + 1
    if uts = ∞ then T.TS $\leftarrow$ tnc   // there is no timestamp adjustment
    else increment timestamp of transactions with timestamp between uts and tnc by 1
        T.TS $\leftarrow$ uts
    endif
endif

Fig. 1. Validation of transaction T

can be eventually committed depends on the results of the reconnection and reintegration process. The data items written by a "pseudo-committed" transaction are visible to subsequent transactions initiated from the same client. They are, however, invisible to other clients and to the server.

2.5 Reconnection and Reintegration

Upon reconnection of a disconnected mobile client, it sends a reconnect message to the server, which acknowledges for reintegration to take place. The mobile client will finish its current transaction and receive validation results from the server, if any. It then sends validation request of its pseudo-committed transactions to the server. Revalidation is conducted at the server, by arranging those pseudo-committed transactions in committed order and the validation process discussed in Sect. 2.3 is performed. If a transaction fails the validation, it will be aborted and all other transactions that read the data items written by the aborted transaction should be aborted. The set of affected transactions can be minimized by computing the set of undesirable transactions for backout [7].

3 Simulation Model

In our simulation, the mobile database environment consists of one database server holding a number of data items. We assume that the setup time for tuning into a broadcast channel and going into doze mode is negligible [10]. A bucket is the smallest logical unit that could be broadcast. All buckets are of the same size for uniformity [6]. A client will submit a request to the server via a point-to-point channel and then wait for reply before it continues processing. The bandwidth of a point-to-point channel is 19.2kps, while the bandwidth of a broadcast channel is 2Mbps. The inter-arrival time for transactions and operations within a transaction is exponentially distributed with a mean of 10 sec. and 10 msec. respectively. Accesses to data items follow a Zipf distribution, modeling the 80/20 rule, or other skewed data access patterns. Each data item is read and written at most once within a transaction; there is no blind write [4].

Each client is installed with a fast SCSI disk, with a bandwidth of 40MBps, which can be in one of three states: read/write, idle, and sleep [6]. We model transitions between disk states based on data access as a measure of energy consumption. There are three states associated with a client mobile data card (MDC): transmit, receive, and sleep [6]; we also model state transition within a MDC for energy consumption. Clients initiate transactions and retrieve data items from appropriate channel or from cache, according to the protocol described in Sect. 2. Server will accept client requests and transmit the required data items. It also performs validation and formal commitment of transactions, installing the updated version.

4 Evaluation

Our simulation model is implemented in C language, using CSIM. Table 1 summarizes the parameter settings in our experiments.

We conduct three representative sets of experiments to evaluate the performance of our protocol. Each set of experiments is repeated five times and the average of each performance metric is taken. The performance metrics include

Parameter	Value
Downlink/uplink channels	10/10
Size of database	1000 items
Item size	50 bytes
Disk idle/write/sleep/spin power	0.65W/0.95W/0.015W/3W
MDC transmit/receive/sleep power	0.4W/0.2W/0.1W
Data access pattern	80/20 rule, Zipf distribution
Read/write operation ratio (ρ)	1, 3, 5, 7
Operations per transaction	4 to 12 uniform
Transactions per client (m)	500
Percentage of read-only transactions	80%
Number of clients (n)	5, 10, 50, 100

Table 1. Parameters for simulated experiments

transaction commit rate, server throughput, and response time experienced by mobile clients. In the first experiment, we study our protocol as compared with the optimistic concurrency control protocol without multi-versioning. As a base case for comparison, we consider an environment using broadcast channels to disseminate data items, while using point-to-point channels only for the purpose of validating update transactions. In the experiment, we vary the number of mobile clients, n, and read/write ratio, ρ, of operations in a transaction and measure the performance metrics. The results are depicted in Fig. 2.

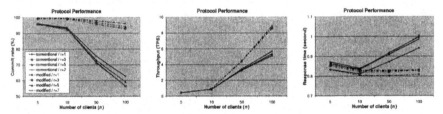

Fig. 2. Performance of concurrency control protocols

The commit rate drops when n increases. This is obvious since the more clients, the more transactions issued and the more potential conflicts which cause abortion of transactions. The commit rate also decreases as ρ decreases, since there would be more write operations in each transaction, leading to more conflicts and thus, more abortions. However, the performance difference due to change in ρ is smaller than one would expect. The throughput increases rapidly with n, since it is the product of number of committed transactions per client (reflected by commit rate) and number of clients. The increase is sublinear, due to reduced commit rate at increased n. Furthermore, the relative performances among protocols and varying values of ρ follow a similar trend with commit rate, with similar reasons.

Our protocol outperforms the conventional optimistic protocol, due to the fact that the use of multiple versions as well as the "consistent" reading from the same broadcast cycle in our protocol improve the transaction commit probability. The performance difference widens as n increases. The response time also follows a similar trend as we expected. The only anomaly we observe is that response time is quite invariant with n under our protocol. Nevertheless, our protocol outperforms the conventional protocol in this aspect. We make a brief measurement on the energy consumption for the conventional protocol and our protocol. Both protocols result in nearly identical disk and mobile data card energy consumption, of values around 7.0W and 1.4W respectively. In our numerous experiments, the energy consumption remains very stable across protocols, though we observe that our protocol yields a smaller variance and fluctuation to the amount of energy consumed across experiments.

Our first experiment has demonstrated a better performance of our protocol over the conventional one. In our second experiment, we thus concentrate on our protocol and contrast its performance in the context of hybrid transaction processing, utilizing both broadcast and point-to-point channels (hybrid), with the base case (base). We vary n and ρ and present results on the three metrics, as shown in Fig. 3.

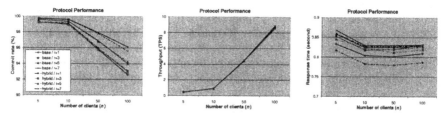

Fig. 3. Performance of hybrid transaction processing

It is clear that hybrid improves the commit rate for small n and ρ, compared with base, but suffers from a decreased commit rate at large n and ρ. This is because the degree of conflict increases with larger value of n, leading to more transactions being aborted for each individual mobile client. Clearly, the throughput increases with n, but at a sublinear manner. Yet, there is no significant difference for the two approaches in terms of throughput. For response time, hybrid performs better than base for all values of n and ρ, due to the reduced reliance by hybrid on the point-to-point channel, shared by all the n clients. The difference in performance is not very large in this experiment, since there are only 1000 data items in the database, resulting in a relatively short broadcast cycle. Other experiments using larger databases and smaller broadcast bandwidth demonstrate much larger differences.

Our final experiment studies the impact of disconnection on mobile clients and the overall impact of reconnection and reintegration on pseudo-committed transactions. We divide each processing cycle into three epochs: a first half, a disconnected period, D, and a second half. Each mobile client will execute about half of its transactions, disconnect for a period D, reconnect and reintegrate, before executing its remaining transactions. We vary D between 100, 200, 300, 400, and 500 seconds, while a transaction processing period is approximately 5000 seconds. The moment when a mobile client starts being disconnected is modeled by a normal distribution, with a mean equal to half the transaction processing time, i.e., about 2500 seconds, and a standard deviation equal to D. This models a spectrum of disconnection events among different clients. We fix n to 50 and consider both base case and hybrid transaction processing as in our second experiment, with ρ equal to 1 and 7. The results are illustrated in Fig. 4.

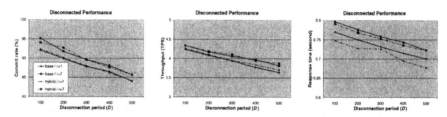

Fig. 4. Performance of disconnected operation

The longer the disconnected period, the worse is the performance in terms of commit rate and throughput. This is obvious since clients will miss more updates from one another, causing a higher degree of conflicts during reintegration. Interestingly, the response time also decreases. This is because the measure of response time is from the initiation of a transaction to the time its fate is known. During a disconnection, a transaction may be aborted or pseudo-committed locally, without having to (and cannot) contact the server for validation. Thus, the response time becomes smaller for those disconnected transactions, bringing down the average measure. We also observe that hybrid performs similarly to base in commit rate, slightly better in throughput, and strictly better in response time. Here, ρ appears to play a more important role in the performance as well.

5   Conclusion and Future Work

In this paper, we proposed to use a combination of broadcast and point-to-point channels to process transactions initiated by mobile clients, by having hot data items disseminated via broadcast channels. We have presented a protocol for hybrid transaction processing in an asymmetric environment. The client cache is adopted to improve performance, especially during disconnection. We have conducted simulated experiments to measure the performance of our protocol, and contrast with that of conventional protocols.

Currently, we assume a fixed hot spot on data items and thus, our broadcast cycle is fixed in length. We are considering adaptivity of data items to be broadcast, leading to varying length of broadcast cycle in different broadcast. Here, the server will have to maintain statistics on data items being accessed and make appropriate decision [6, 13]. A combined usage of broadcast and point-to-point channels implies that there should be an efficient channel allocation algorithm to cater for both types of data access requirements, while producing an optimal overall performance. An adaptive channel allocation algorithm is required for switching between channel types [8, 16]. We would also like to improve the access and dissemination of broadcast data items [2] and in the event of disconnected operation, the utilization of cached data items [5]. Finally, it might be beneficial to maintain a dynamic number of versions for hot data items to improve the commit rate of read-only transactions.

References

1. Acharya, S., Alonso, R., Franklin, M., Zdonik, S.: Broadcast disks: Data management for asymmetric communication environments. In *Proceedings of ACM SIGMOD Conference*, pages 199–210, 1995.
2. Acharya, S., Franklin, M., Zdonik, S.: Balancing push and pull for data broadcast. In *Proceedings of ACM SIGMOD Conference*, pages 183–194, 1997.
3. Barbara, D.: Certification reports: Supporting transactions in wireless systems. In *Proceedings of International Conference on Distributed Computing Systems*, pages 466–473. 1997.
4. Bernstein, P.A., Hadzilacos, V., Goodman, N.: *Concurrency Control and Recovery in Database Systems*. Addison-Wesley, Reading, Massachusetts, 1987.
5. Chan, B.Y.L., Leong, H.V., Si, A., Wong, K.F.: MODEC: A multi-granularity mobile object-oriented database caching mechanism, prototype and performance. *Journal of Distributed and Parallel Databases*, 7(3):343–372, 1999.
6. Datta, A., VanderMeer, D.E., Celik, A., Kumar, V.: Adaptive broadcast protocols to support efficient and energy conserving retrieval from databases in mobile computing environment. *ACM Transactions on Database Systems*, 24(1):1–79, 1999.
7. Davidson, S.B.: Optimism and consistency in partitioned distributed database systems. *ACM Transactions on Database Systems*, 9(3):456–481, 1984.
8. Hu, Q., Lee, D.L., Lee, W.C.: Optimal channel allocation for data dissemination in mobile computing environment. In *Proceedings of International Conference on Distributed Computing Systems*, pages 480–487. 1998.
9. Imielinski, T., Badrinath, B.R.: Mobile wireless computing: Challenges in data management. *Communications of the ACM*, 37(10):18–28, 1994.
10. Imielinski, T., Vishwanathan, S., Badrinath, B.R.: Data on air: Organization and access. *IEEE Trans. on Knowledge and Data Engineering*, 9(3):353–372, 1997.
11. Kistler, J., Satyanarayanan, M.: Disconnected operation in the Coda file system. *ACM Transactions on Computer Systems*, 10(1):3–25, 1992.
12. Lee, K.C.K., Leong, H.V., Si, A.: Semantic query caching in a mobile environment. *ACM Mobile Computing and Communications Review*, 3(2):28–36, 1999.
13. Leong, H.V., Si, A.: Database caching over the air-storage. *The Computer Journal*, 40(7):401–415, 1997.
14. Pitoura, E., Chrysanthis, P.K.: Scalable processing of read-only transactions in broadcast push. In *Proceedings of International Conference on Distributed Computing Systems*, pages 432–439. 1999.
15. Pitoura, E., Chrysanthis, P.K.: Exploiting versions for handling updates in broadcast disks. In *Proceedings of International Conference on Very Large Data Bases*, pages 114–125. 1999.
16. Prakash, R., Shivaratri, N.G., Singhal, M.: Distributed dynamic channel allocation for mobile computing. In *Proceedings of ACM Annual Symposium on Principles of Distributed Computing*, pages 47–56. 1995.
17. Shanmugasundaram, J., Nithrakashyap, A., Sivasankaran, R., Ramamritham, K.: Efficiency concurrency control for broadcast environments. In *Proceedings of ACM SIGMOD Conference*, pages 85–96, 1999.

Speculative Lock Management to Increase Concurrency in Mobile Environments

P. Krishna Reddy and Masaru Kitsuregawa

Institute of Industrial Science, The University of Tokyo
7-22-1, Roppongi, Minato-ku, Tokyo 106, Japan
{reddy, kitsure}@tkl.iis.u-tokyo.ac.jp

Abstract. Since mobile transactions are long-lived, in case of a conflict, the waiting transactions are either blocked for longer durations or aborted if two-phase locking is employed for concurrency control. In this paper, we propose mobile speculative locking (MSL) protocol to reduce the blocking of transactions. MSL allows a transaction to release a lock on a data object whenever it produces corresponding after-image value. By accessing both before- and after-images, the waiting transaction carries out speculative executions at the mobile host. Before the end of commit processing, the transaction that has carried out speculative executions retains appropriate execution based on the termination decisions of preceding transactions. The MSL approach requires extra resources at the mobile host to carry out speculative executions. Since mobile host is operated by single user, we assume that it can support a reasonable number of speculative executions. Through analysis it has been shown that MSL increases concurrency with limited resources available at mobile host.

1 Introduction

Technological advances in cellular communications, wireless LAN and satellite services have led to the emergence of the mobile computing environments. It is expected that in near future, millions of users will carry a portable computer and communication devices that uses wireless connection to access world wide information network [9]. To support mobile computing, the transaction processing models should accommodate the limitations of mobile computing, such as unreliable communication, limited battery life, low bandwidth communication, and reduced storage capacity. Operations on shared data must ensure correctness of transaction executed on both stationary and mobile hosts. Also, the blocking of transaction execution on either the stationary or mobile host must be minimized to reduce communication cost and increase concurrency.

Due to the fact that mobile transactions are long-lived, in case of a conflict, the waiting transactions are either blocked for longer durations or aborted if two-phase locking is employed for concurrency control. In this paper we propose mobile speculative locking (MSL) approach to increase concurrency. In the M-SL approach, a transaction releases locks on a data object whenever it writes

H.V. Leong et al. (Eds.), MDA'99, LNCS 1748, pp. 82–96, 1999.

corresponding data object. The waiting transaction reads both before- and after-image and carries out speculative executions. However, the transaction which has carried out speculative executions can commit only after termination of preceding transactions. Up on termination of preceding transactions, it selects appropriate execution based on the termination decisions of preceding transactions. In this way, by reducing blocking, MSL increases concurrency without violating serializability criteria.

In this paper, we explain MSL for a generalized case. That is, a transaction which has carried out speculative executions in turn releases the locks on data objects whenever it produces corresponding after-images. As such, there is no limitation on the number of levels of speculation but this number depends on the system's resources, such as the size of main memory and processing power of mobile host. Due to reliance on parallel computation to carry out speculative executions, the MSL approach requires both computing resources at mobile host (MH) to support speculative executions. Also, the increase in concurrency depends on the number of speculative executions supported by MH.

We assume that mobile host could be equipped with extra processing power and main memory to carry out speculative executions. This assumption is valid due to the fact that MH is operated by single user and all available resources at MH are at his disposal. In near future, even though, there exists no hope to increase battery life, processing power and main memory will be increased to support multimedia applications. We have proposed MSL for non-interactive transactions. Also, the speculative executions are transparent to the user.

The remaining paper is organized as follows. In the next section we present the review of previous research. In section 3 we explain mobile database system model. In section 4, we present MSL protocol and discuss modification of MSL according to limited resources available at MH. In section 5, we present case example using MSL approach. In section 6, thrpugh analysis we show how MSL increases concurrency. In section 7, we discuss message and processing overheads of MSL. Finally, the last section consists of summary and conclusions.

2   Related Work

In this section we first review some research papers related to mobile transaction management. We also review the research work related to transaction management in conventional database systems, which has influenced this research.

In [2, 9] some of the problems involved in supporting transaction services and distributed data management in a mobile environment have been identified. The management of distributed data has been identified in [9] as a research area on which the mobility of host has a large impact. A mobile transaction model that captures both data and movement behavior is proposed in [7]. In [17] semantic database models have been extended for mobile computing environment to increase concurrency by exploiting commutative operations. In the clustering model [14], the database is divided into clusters and a mobile transaction is decomposed into a set of weak and strict operations, The decomposition is done based on the consistency requirement. The read and write operations are also classified as weak and strict. The weak operations are allowed to access only data elements belong to the same cluster, where as strict operations are allowed to database wide access.

We now review some concurrency control techniques and performance studies which are based on early release of lock and speculation.

In [4] speculation has been employed to increase the transaction processing performance for real-time centralized environments that employ optimistic algorithms for concurrency control. In [3], a branching transaction model has been proposed for parallel database systems where a transaction follows alternative paths of execution in case of a conflict. However, this paper has not discussed the case of distribution. Also, the operation in limited resource environments is not analyzed. In [10] a proclamation-based model is proposed for cooperative environments in which a cooperative transaction proclaims a set of values, one of which a transaction promises to write if it commits. The waiting transactions can access these proclaimed values and carry out multiple executions. This approach is mainly aimed at cooperative environments such as design databases and software engineering. In [16] the pre-write operation is introduced to increase concurrency in a nested transaction processing environment.However, it is assumed that once the sub-transaction pre-writes the value, it will not abort in future. In [13], we have proposed concurrency control approach for distributed database systems in that a transaction releases the locks at the end of the execution and carries out two-phase commit processing. Through simulation study it has been shown that speculation considerably improves the performance in case of higher conflicts and longer transmission times.

In MSL a transction releases locks before execution. We propose MSL as an alternative to two-phase locking with the assumption that the MH could support a reasonable number of speculative executions. Also, we analyze how MSL could increase concurrency with limited resources available at the MH.

3   Mobile Transaction Model

Data objects are represented by $X, Y, \ldots, Z$. Transactions are represented by $T_i, T_j, \ldots$; and object servers are represented by $S_i, S_j, \ldots$; where, $i, j, \ldots$ are integer values. The data objects are stored at object servers connected by a computer network. Each data object is stored at one fixed server only (no replication).

Figure 1 presents a general mobile database system (MDBS) model similar to those described in [9] for mobile computing systems. Each mobile support station (MSS) has a coordinator which receives transaction's operations from mobile tranaction hosts (MTHs or simply MHs) and monitors their execution in database servers within the fixed networks. A database server is to support basic transaction operations such as read, write, prepare, commit, and abort. Transaction operations are submitted by a MH to the coordinator in its MSS, which in turn sends them to the distributed database servers within the fixed networks for execution. Similarly, for a commit operation, the coordinator monitors the execution of two-phase commit (2PC) [6] protocol over all the servers involved in the execution of the transaction.

In this paper we assume that operations of a transaction may be submitted in multiple request messages rather than submitting an entire transaction as a single request message to the coordinator [11]. A submission unit thus consists of one operation (e.g., read) or a group of operations. The mobile host interactively submits the operations of a transaction to its coordinator. The subsequent operation can be submitted only after those previous have been executed and results returned from the coordinator. This approach involves multiple coordinators because of the mobility of MH.

Knowledge of after-image : Normally, a transaction copies data objects through read operations into private working space and issues a series of update operations [6]. We assume that for T_i and data object X, $w_i[X]$ operation is issued whenever T_i completes work with the data object. This assumption is also adopted in [1, 15].

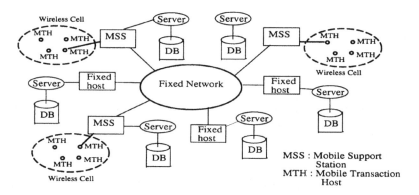

Fig. 1. Mobile database system model

4   Mobile Speculative Locking

4.1   Lock Compatibility Matrix

In the MSL approach, we employ three types of locks: R, EW-, and SPW-locks. A transaction requests R-lock to read, and EW-lock to read and write. Transactions request only R- and EW-locks. When a transaction obtains an EW-lock, the EW-lock is changed to SPW-lock only after inclusion of after-images to the respective trees (tree will be explained later) of data objects.

The lock compatibility matrix for MSL is shown in Figure 2. It can be observed that at any point in time, only one transaction holds an EW-lock on a data object. However, multiple transactions can hold R- and SPW-locks simultaneously. The MSL approach ensure consistency by forming *commit dependency* among transactions. If T_i forms a *commit dependency* with T_j, T_i is committed only after termination of T_j. In Figure 2 the entry *sp_yes* (speculatively yes) indicates that the requesting transaction carries out speculative executions and forms a *commit dependency* with lock-holding transaction.

Lock requested by T_j	Lock held by T_i		
	R	EW	SPW
R	yes	no	sp_yes
EW	sp_yes	no	sp_yes

Fig. 2. Lock compatibility matrix of MSL

4.2 MSL Algorithm

We first explain the main data structures used to present the MSL approach[1] assuming unlimited resources at MH. Next, we explain the modification assuming limited resources.

- $tree_X$:   We employ a tree data structure to organize the uncommitted versions of a data object produced by speculative executions. The notation $X_q(q \geq 1)$ is used to represent the qth version of X. For a data object X, its object tree is denoted by $tree_X$. It is a tree with committed version as the root and uncommitted versions as the rest of the nodes.
- $depend\_set_{ij}$: $depend\_set_{ij}$ is a set of transactions from which T_i has formed commit dependencies at the object server S_j. It is maintained at S_j.
- $queue_X$: queue of X, which is maintained for each X at its resident site. Each element in $queue_X$ is a two tuple $< indentity\ of\ lock\ request,\ status >$. In this status takes two values: waiting and not-waiting.

The server where the object resides is referred to as an object server. We use the notation T_{im} to represent the mth ($m \geq 1$) speculative execution of T_i. Each data object X is organized as a tree with X_1 as a root at the server. Also, read-set (RS) and write-set (WS) of T_{im} are represented by RS_{im} and WS_{im} respectively. Lock request of T_i for X that resides at S_j is denoted by $LR_{ij}(X)$.

1. **Lock acquisition**
   Suppose MH starts T_i. It submits lock request $LR_{ij}[X]$ to the nearest coordinator which in turn forwards to S_j. If this is T_i's fiirst request to S_j, $depend\_set_{ij}$ is initialized to ϕ. If LR_{ij} is in conflict with preceding LR_{kj}

---

[1] Proof of correctness is not provided due to space limitation. However, one could easily prove that all MSL histories are serializable[6].

which is already waiting for X in $queue_X$, the tuple $< LR_{ij}, waiting >$ is enqueued to $queue_X$. Otherwise, if preceding LR_{kj} has accessed X, and holds R/SPW-lock the steps given below are followed. If $LR_{ij}(X)$ forms a commit dependency with $LR_k(X)$, R/EW-lock is granted to $LR_i(X)$, $depend\_set_{ij}$ updated as $depend\_set_{ij} \cup \{T_k\}$. The status is changed to "not-waiting". Next, $tree_X$ with lock reply message is sent to MH.

2. **Speculative processing**

   At MH, on receiving a lock reply message, T_i is processed as follows.

   – Suppose $tree_X$ contains 'n', versions and T_i is carrying out 'm' speculative executions. Then, each T_{iq} (q=1 ... m), splits into 'n' speculative executions (one for each version of $tree_X$).

   – During execution, whenever a transaction issues a write operation on X, corresponding X's after-images of all speculative executions are sent to the nearest coordinator, which in turn forwards to X's object server. At the object site, each new version of X is included as a child to the corresponding read version node of $tree_X$ and the EW-lock is converted to the SPW-lock. A waiting lock request is granted a R/EW-lock.

3. **2PC processing**

   On completion of execution, T_i (it's MH) sends RS and WS of all its speculative executions to nearest coordinator and submits a commit request. The coordinator starts 2PC and sends PREPARE messages to the object sites and the coordinators of the cells in which T_i has visited[2].

   At the object server S_j if $depend\_set_{ij} = \phi$, it sends a $VOTE\_COMMIT$ message. Otherwise, it waits for the termination of all transactions in $depend\_set_{ij}$. When $depend\_set_{ij} = \phi$, for all the data objects accessed by T_i at S_j, the identities of respective root node values are sent with $VOTE\_COMMIT$ message.

   After receiving $VOTE\_COMMIT$ messages from all the participants, the coordinator selects the speculative execution to be confirmed and sends identity of after-images with $GLOBAL\_COMMIT$ message. Otherwise, it sends a $GLOBAL\_ABORT$ message. Also, it informs the MH about the decision.

   If a participant receives the a $GLOBAL\_COMMIT$ message, the tree of corresponding data object is replaced by the subtree with the selected after-image as the root node. Otherwise, if a participant receives a $GLOBAL\_ABORT$ message, the after-images included by the transaction are deleted along with subtree. When a transaction is terminated (aborts or commits), the termination information is broadcasted to all the concerned MHs and object servers. On receiving this information, the waiting transactions drop speculative executions carried out by reading dirty/invalid data. At the object servers, the trees and $depend\_sets$ are updated similar to the case of $GLOBAL\_ABORT$ or $GLOBAL\_COMMIT$.

---

[2] In mobile environment, the coordinators of the cells visited by T_1 can issue an abort command independently. To ensure consistency we have to involve coordinators in addition to object servers during 2PC.

4. Deadlocks

When a LR of a transaction conflicts, deadlock detection [12] needs to be initiated. Further, even if a transaction forms commit dependency, may form a commit dependency cycle. Note that a cycle in the dependency graph may involve both commit dependency and wait-for edges. The process of checking deadlocks and commit dependency cycles can be achieved using single wait-for-graph (WFG) [5].

4.3   Limited Resources

We now explain the methods to keep number of executions within the manageable limit according to resources available at MH. The number of speculative executions carried out by a transaction could be controlled with two variables : *versions_limit* and *executions_limit*.

- *Versions_limit*: For a data object, the *versions_limit* variable limits the maximum number of versions in the tree of a data object.
- *Executions_limit*: Suppose the amount of memory to carry out a single execution is one unit. The *executions_limit* variable is set to maximum number of speculative executions that can be carried out by the MH.

Modification of MSL : We assume that we know the *executions_limit* value, calculated from the main memory size of MH and average memory required to execute a transaction. Suppose, a transaction is carrying out N ($N \geq 1$) speculative executions. The *versions_limit* value for the subsequent lock request is calculated as $\lfloor \frac{executions\_limit}{N} \rfloor$. This information is sent along with the lock request to the object site. If EW-lock is available and if number of versions is less than or equal to *versions_limit* value, the tree of corresponding data objects is shifted to MH. Otherwise, if number of versions in the tree of a data object is greater than *versions_limit*, the lock request is made to wait. That is, the lock request waits for the termination of earlier transactions. On termination of earlier transactions (the tree is modified) if the number of nodes is less or equal to the *versions_limit*, the lock is granted to the waiting lock request. Also, at the MH, if it is unable to carry out required number of speculative executions, it waits for the termination information of earlier transactions.

5   Case Example

Consider that data objects W, X, Y, and Z are stored at object servers S_1, S_2, S_3 and S_4 respectively. Suppose, the following three transactions are issued by different mobile hosts. These are, T_1 : $r_1[X]\ w_1[X]\ r_1[Y]\ w_1[Y]$; T_2 : $r_2[X]\ w_2[X]\ r_2[Z]\ w_2[Z]$; and T_3 : $r_3[X]\ w_3[X]\ r_3[W]\ w_3[W]$. (For a T_i, $r_i[X]$ and $w_i[X]$ denote read and write operations on X.)

Using the MSL approach, these transactions can be processed as follows. For this example, we assume that transactions request only EW-locks. We employ

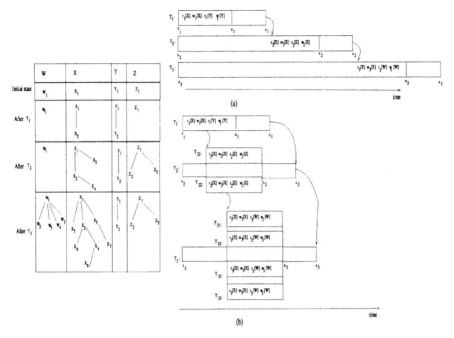

Fig. 3. Depiction of three growth

Fig. 4. Processing:(a)with 2PL (b)with MSL

the nested parentheses notation to represent object trees. Initially W, X, Y and Z are organized as trees at respective object servers as W_1, X_1, Y_1, and Z_1 as root nodes respectively.

Consider that T_1 first obtains EW-lock on X (by sending LR to S_2 through nearest coordinator) and starts execution T_{11} by reading X_1. Whenever it produces after-image X_2, it sends X_2 to S_2. X_2 is included in $tree_X$ as $(X_1(X_2))$ and lock on X is changed to SPW lock. Next it requests a lock on Y. Whenever it produces after-image Y_2, it sends Y_2 to S_3. At S_3, Y_2 is included in $tree_Y$ as $(Y_1(Y_2))$ and lock on Y is changed to SPW lock.

When T_1 changes lock on X from EW to SPW-lock, T_2 obtains an EW lock on X. At S_2, T_1 is included in $depend\_set_{22}$. Next, T_2 carries out two speculative executions: T_{21} and T_{22}. T_{21} reads X_1 produces X_3. T_{22} reads X_2 produces X_4. Afterwards, both X_3 and X_4 are sent to S_2, which are included in $tree_X$ as $(X_1(X_2(X_4), X_3))$. Lock on X is converted to SPW lock. Next T_2 obtains EW lock on $tree_Z$. Both T_{21} and T_{22} access Z_1 and produce Z_2 and Z_3 respectively. Next, both Z_2 and Z_3 are sent to S_4 which are included in $tree_Z$ as $(Z_1(Z_2, Z_3))$. Next, lock on Z is changed to SPW-lock.

Similarly, T_3 obtains EW-lock on X and carries out four executions. The growth of the object trees of W, X, Y, and Z are depicted in Figure 3. Figure 4(a) depicts processing with 2PL and Figure 4(b) depicts processing with MSL.

(In Figure 4, s_i, e_i and c_i denote starting of execution, completion of execution and completion of commit of T_i respectively. Also, an arrow from from a to b, indicates b happens after a)

When a transaction completes execution, commit processing is carried out by nearest coordinator. The coordinator sends PREPARE messages to the participating sites. In case of T_2, as $T_1 \in depend\_set_{22}$ at S_2, $VOTE\_COMMIT$ message is sent only after termination of T_1. Similarly, in case of T_3, PREPARE messages are sent to both S_2 and S_1. At S_2, because both $T_1, T_2 \in depend\_set_{32}$, $VOTE\_COMMIT$ message is sent only after the termination of both T_1 and T_2. If T_1 is committed, the trees of X and Y are updated as X_2 and Y_2 as roots. Otherwise, if T_1 is aborted, the nodes X_2 and Y_2 are removed along with subtrees. In both cases, after updating object trees the identity of T_1 is removed from $depend\_set$ of other lock requests.

6   Concurrency Analysis

In MSL approach, speculative executions of transaction depends on its speculation level and number of data objects it conflicts with other transactions. We first define the term *speculation level* which is used to quantify the parallelism that could be achieved using SL.

Definition 1 (Speculation level):   For T_j, the speculation level is denoted by ρ_j. If T_j executes without conflict, $\rho_j = 0$. Let T_j speculatively reads a set of data objects, say, *spec_set* updated by 'n' transactions. Each $X \in spec\_set$ is updated by some T_k, at speculation level ρ_k. Let ρ_{max} be the maximum of all ρ_k, where T_k has updated a data object in *spec_set*. Then, $\rho_j = (\rho_{max} + 1)$.

Now we derive relationship between speculative executions of a transaction and speculation level.

Let T_i conflicts on m $(m \geq 0)$ data objects with other transactions. When T_i obtains lock on first data object with v_1 nodes in its tree, it carries out v_1 executions. When it accesses the second object having v_2 nodes in its tree, each one of the v_1 executions carries out v_2 executions. Following this, after accessing all m objects, the total number of speculative executions carried out by $T_i = \prod_{k=1}^{m} v_k$.

Note that, if a transaction has no conflict with other transaction on the kth data object, v_k is one. Otherwise, if a transaction obtains the lock on k^{th} data object in speculative mode (some other transaction has updated the object tree), $v_k > 1$. For the sake of simplicity, let c be the mean of number of data objects that a transaction conflicts, ρ be the mean speculation level. Also, let v_ρ be the mean of number of versions in the tree of a data object and N_ρ be number of executions at level ρ. Then,

$$N_\rho = v_\rho^c \tag{1}$$

Suppose, the number of new versions generated at level ρ for each data object accessed by a transaction be new_ρ. The number of versions at the next level $v_{\rho+1}$ is given below.

$$v_{\rho+1} = v_\rho + new_\rho \quad and \quad v_0 = 1 \tag{2}$$

For the sake of simplicity we make two worst-case assumptions. First, we assume that a transaction requests only write locks and releases these locks after completing execution. And second when a transaction carries out N_ρ executions, N_ρ distinct versions are included to the tree of each data object it accessed after its execution. Then,

$$new_\rho = N_\rho \tag{3}$$

Substiting new_ρ in Equation 2

$$v_{\rho+1} = v_\rho + N_\rho \quad where \quad v_0 = 1, N_0 = 1 \tag{4}$$

Object dependency

As per our assumptions when a transaction carries out N executions, N new versions are included in the tree of each data object. This is true if each speculative execution produces a distinct after-image for data object read by it. However, for transactions in real applications (such as banking and airline reservations) each speculative execution may not produce a distinct after-image value. If computation on an an object X is independent of other data object values it has accessed, then each speculative execution does not produce new version. We call this property **object-dependency**.

Definition 2 (depend(X,Y)) Given transaction T_i that accesses n objects ($n \geq 2$). Let X and Y be any data objects accessed by T_i. depend(X,Y)=true, if $w_i[Y]$ is an arbitrary function of $r_i[X]$. Otherwise, depend(X,Y)=false. We assume that for a data object X, $depend(X, X) = true$ always holds.

We can extend Definition 2, in a transitive fashion. For instance, if depend(X,Y)=true and depend(Y,Z)=true, then depend(X.Z)=true.

When a transaction conflicts on c data objects, each data object Y may form distinct *depend* relationship with other objects. However, for our analysis, we are interested in that $depend(X,Y)$ relationship in which Y transitively forms with X, through maximum number of data objects.

Definition 3 (dependency_size$_y$) : Given transaction T_i that conflicts with other transactions on c objects. Let X be any data object accessed by T_i. Let Y forms a depend(X,Y), transitively, through maximum number of data objects. $dependency\_size_Y$ is the number of data objects in such transitive path (including Y).

Definition 4 (object dependency factor (odf)): Among *dependency_size* of all the data objects accessed by T_i let p be the maximum *dependency_size*. Then,

$$odf = \frac{p}{c} \quad if \ 1 \le p \le c, \ c \ge 1, \tag{5}$$

Let α be mean odf. With object dependency factor, we rewrite new_ρ as follows.

$$new_\rho = v_\rho^{\alpha \times c} \quad for \quad \frac{1}{c} \le \alpha \le 1 \tag{6}$$

Substituting new_ρ in Equation 2

$$v_{\rho+1} = v_\rho + v_\rho^{\alpha \times c} \quad where \quad \frac{1}{c} \le \alpha \le 1 \ and \ v_0 = 1 \tag{7}$$

6.1   Performance Discussion

Based on preceding analysis we discuss how MSL could increase concurreny at different data contention levels and resources available at MH.

A **Single conflict (Hot spots)** Let N represents number of memory units (where each execution requires one memory unit) available at MH. It can be observed that, in database environments where the majority of transactions conflict on a single data object, MSL considerably increases the concurrency with manageable value of N. From Equations 1 and 7, with c=1, the relationship between ρ and N_ρ is, $N_\rho = 2^\rho$. Therefore, $\rho = \log N_\rho$. From Figure 5(a), for instance, it can be observed that by supporting eight speculative executions for a transaction at MH (i.e., with N=8 and c=1), concurrency could be increased up to three speculation levels.

B **Multiple conflicts (long transactions)**
With $\alpha = 1$ and ρ=1, the relationship between c and N_1 is, $N_1 = 2^c$. When a transction encountters multiple conflicts, if we support 2^c speculative executions, concurrency could be increased up to one speculation level. Figure 6(a) depicts the scenario. With $\alpha = 1$ and $\rho = 2$, the relationship between c and N_2 is, $N_2 = (2 + 2^c)^c$. We can observe that 2-level speculation is manageable to an extent at higher N values. From Figure 5(a), it can be observed that with c=2, concurrency can be increased up to 2-levels at N=36. But, at multiple conflicts ($c > 2$) and higher speculation levels ($\rho > 2$), the value of N explodes.

However, at higher speculation levels, α controls the explosion of speculative executions. From Figure 5(b), when c=2, and $\alpha = \frac{1}{c}$), concurrency can be increased up to 2 speculation levels with N=16. In this way in multiple conflict environments, MSL increases concurrency by exploiting object dependency property among data objects.

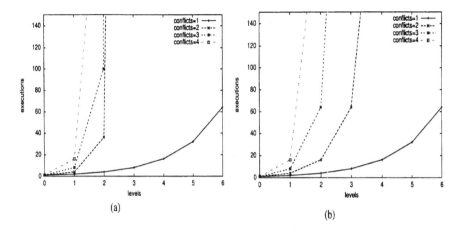

Fig. 5. Number of levels versus speculative executions (a) $\alpha = 1$ (b) $\alpha = \frac{1}{c}$.

The preceding analysis is carried out with simplified (worst case) assumptions. However, in real environments a transaction reads more data objects and writes few. When a transaction reads, no new versions are included to corresponding data object tree. Also, MSL updates the data object tree whenever a transaction produces corresponding after-image. Therefore the number of versions included to a data object is equal to number of executions at that instant of time. These factors decrease number of versions included to the tree of a data object, which further increase the concurrency.

7 Message and Processing Issues

7.1 Message Overheads

- **Updating trees**
  In MSL, a transaction updates the data object tree whenever it issues corresponding write operation. In mobile environment, the link between MH to fixed network is weak (the bandwidth is less). This message contains after images of all speculative executions of corresponding object. The size of the message depends on number of distinct after-images produced by a transaction. We can employ lossless compression techniques to reduce the message size. Even though this incurs little overhead, the performance can be significantly improved as conflicting transactions are processed in parallel.
- **Termination messages**
  In MSL, whenever a transaction terminates, this information is to be sent to all MHs and coordinators that have carried out speculative executions based on the values written by terminated transactions. In mobile environment, this overhead is acceptable, since object servers and coordinators of cells

are attached to high speed network. Also, broadcasting capability is used to inform concerned MHs about termination decision.

- **Commit processing**
  Number of messages during commit processing is not increased. The coordinator is attached to high speed network which performs the commit processing and informs the final decision to MH.

7.2   Processing Overheads

- **Main memory at object servers**
  MSL requires extra main memory at object servers to maintain trees for objects.
- **Resources at MH**
  Note that, in mobile environment, MH is operated by single user. All the processing resources available at MH are at his disposal. Even with available RAM and processing power with current technology, MH could easily support a reasonable number (around 10 or more) of speculative executions which reduces the overall blocking of transactions, significantly.
- **Battery life**
  Using MSL, battery power can be saved considerably as MSL reduces blocking time significantly. Because, MSL allows the MH to get the data object values early as compared to traditional 2PL.
- **No extra disk cost**
  Also, this approach involves no additional disk write cost for recovery. The coordinator force-writes corresponding after-images only after selecting the appropriate after-image value. The participant force-writes corresponding after-images only after receiving $GLOBAL\_COMMIT$ message from the coordinator. Therefore, even though a transaction produces number of versions behalf of a data object during its execution, force-writes only one version, after selecting the speculative execution to be confirmed. Therefore, no extra disk write cost is involved.

8   Summary and Conclusions

In this paper we have proposed concurrency control approach based on speculation for mobile database environments. In the proposed MSL approach, a transaction releases a lock on a data object whenever it produces corresponding after-image value during its execution. By this the blocking could be reduced, with out violating serializability criteria and cascading aborts. To increase concurrency, MSL requires extra memory and processing resources at MH to support speculative executions. MSL trades extra resources of MH to increase concurrency. Through analysis it has been shown that MSL increases concurrency significantly even we support reasonable number of speculative executions at MH.

As a part of future work, we will evaluate performance through simulation experiments and extend MSL to interactive environments. In this paper we have considered flat transaction model to present MSL. However, it has been reported [7] that nested transaction model naturally fits in mobile environments. We will extend MSL to nested transaction processing environments. If we employ speculation, sub-transactions release their locks before their ancestor transaction's commit. This allows other sub-transactions to acquire required locks earlier which increases parallelism.

Recent advances in communication technology improves the transmission bandwidth significantly. However, we are still suffering from message latency and it will be very difficult to resolve this problem in near future. The latency problem is amplified when long communication delays are common in mobile environments due to mobility factor. MSL plays an important role under such latency sensitive mobile database applications.

Acknowledgments

This work is partially supported by Grant-in-Aid for Creative Basic Research # 09N-P1401: "Research on Multi-media Mediation Mechanism for Realization of Human-oriented Information Environments" by the Ministry of Education, Science, Sports and Culture, Japan and Japan Society for the Promotion of Science.

References

1. D.Agrawal, A.El Abbadi, and A.E.Lang, The performance of protocols based on locks with ordered sharing, *IEEE Transactions on Knowledge and Data Engineering*, vol.6, no.5, October 1994, pp. 805-818.
2. R.Alonso and H.F.Korth, Database system issues in nomadic computing, Proc. of the 1993 ACM SIGMOD, pp. 388-392, 1993.
3. A.Burger and P.Thanisch, Branching transactions: a transaction model for parallel database systems, *Lecture Notes in Computer Science 826*.
4. Azer Bestavros and Spyridon Braoudakis, Value-cognizant speculative concurrency control, proc. of the 21th VLDB Conference, pp.122-133, 1995.
5. B.R. Badrinath and K.Ramamritam, Semantics based concurrency control: Beyond commutativity, *ACM Transactions on Database Systems*, vol.17, no.1, March 1992, pp. 163-199.
6. P.A.Bernstein, V.Hadzilacos and N.Goodman, *Concurrency control and recovery in database systems*(Addison-Wesley, 1987).
7. M.H.Dunham, A.Helal, and S.Balakrishnan, A mobile transaction model that captures both the data and movement behavior, Mobile Networks and Applications, vol. 2, pp. 147-162, 1997.
8. A.K.Elmagarmid, A survey of distributed deadlock detection algorithms, *ACM SIGMOD RECORDS*, 15(3), September 1986, pp. 37-45.
9. T. Imielinksi, and B.R.Badrinath, Wireless mobile computing: Challenges in Data Management, Communications of ACM, 37(10), October 1994.
10. H.V.Jagadish, and O.Shmueli, A proclamation-based model for cooperation transactions, proceedings of the 18th VLDB Conference, Canada, 1992.
11. J.Jing, O.Bukhres, and A.Elmagarmid, Distributed lock management for mobile transactions, in proceedings of 15th International Conference on Distributed Computing Systems, pp. 118-125, June 1995.

12. E.Knapp. Deadlock detection in distributed databases, *ACM Computing Surveys*, 19(4), December 1987, pp.303-328.
13. P.Krishna Reddy and Masaru Kitsuregawa, Improving performance in distributed database systems using speculative transaction processing, *in proceedings of The Second European Parallel and Distributed Systems conference (Euro-PDS'98)*, 1998, Vienna, Austria.
14. E.Pitoura, and B.Bhargawa, Maintaining Consistency of Data in Mobile Computing Environments, in proceedings of 15th International Conference on Distributed Computing Systems, pp. 404-413, June 1995.
15. K.Salem, H.Garciamolina and J.Shands, *Altruistic locking*, ACM Transactions on Database Systems, vol. 19, no.1, March 1994, pp. 117-165.
16. S.K.Madira, S.N.Maheswari, B.Chandra and Bharat Bhargawa, Crash Recovery algorithm in open and safe nested transaction model, *Lecture Notes in Computer Science*, vol. 1308, Springer-Verlag, 1997, pp. 440-451.
17. G.D.Walborn, and P.K.Chrysanthis, Supporting semantics-based transaction processing in mobile database applications, in proceedings of 14th IEEE Symposium on Reliable Distributed Systems, pp. 31-40, 1995.

Optimistic Concurrency Control in Broadcast Environments: Looking Forward at the Server and Backward at the Clients

Victor C. S. Lee and Kwok-wa Lam

Department of Computer Science, City University of Hong Kong
csvlee@cityu.edu.hk

Abstract. In data broadcast environments, the limited bandwidth of the upstream communication channel from the mobile clients to the server bars the application of conventional concurrency control protocols. In this paper, we propose a new variant of the optimistic concurrency control protocol that is suitable for the broadcast environments. In this protocol, read-only mobile transactions can be processed locally at the mobile clients. Only update transactions are sent to the server for final validation. These update transactions will have a better chance of commitment because they have gone through partial validation at the mobile clients. This protocol, while using less control information to process transactions at the mobile clients, provides autonomy between the mobile clients and the server with minimum upstream communication, which are desirable features to the scalability of applications running in broadcast environments.

1 Introduction

Data broadcast environments pose a number of challenging issues on transaction processing. In these environments, the server may have a relatively high bandwidth broadcast capability while the bandwidth for mobile clients to the server is very limited. Such asymmetric communication environments [1] render the conventional transaction processing mechanisms, which require bi-directional communication between server and clients, inapplicable since the time required for message passing may be intolerably long. Another issue is that data broadcast over the air is monetarily expensive [7] as the bandwidth will continue to be a scarce resource [15]. The prolonged communication time may be too costly. Furthermore, the large population size of mobile clients may overload the server when they asynchronously submit transactions to the server for processing. Therefore, one of the design objectives of our protocol is to minimize the use of upstream bandwidth.

In optimistic concurrency control protocols, transactions are allowed to execute unhindered until they reach their commit point, at which time they are validated [10]. This approach to processing transactions can suit well to the asymmetric communication bandwidth in broadcast environments. The server

H.V. Leong et al. (Eds.), MDA'99, LNCS 1748, pp. 97–106, 1999.
© Springer-Verlag Berlin Heidelberg 1999

can use the large downstream bandwidth to broadcast data objects to transactions at the mobile clients such that part of the transaction execution can be processed locally without sending every request to the server. Upon completion of a transaction at the mobile client, the transaction is sent back to the server for validation. This basic application of optimistic approach may help to solve the limited upstream communication bandwidth. However, this approach has at least two weaknesses in broadcast environments. Firstly, some mobile transactions, which are destined to restart because of data conflicts with committed transactions at the server, can be executed to the end. This useless processing wastes resources at the mobile clients. Secondly, sending these transactions, which will eventually be restarted at the server, wastes the limited upstream communication bandwidth.

Data management in wireless environments receives a lot of attention in these few years [2], [4], [8], [9], [11], [13], [15], [17], [18]. However, there are only a few studies on transaction processing and nearly all of them are focused on the processing of read-only transactions. In our protocol, both read-only and update transactions at the mobile clients are catered.

2 Issues of Transaction Processing in Broadcast Environments

2.1 Serializability

Existing concurrency control protocols for transaction processing are not suitable for broadcast environments due to the asymmetric communication bandwidth. Hence, recent work [18], [19] relaxed the strictness of serializability and proposed some concurrency control protocols based on the relaxed consistency requirements for broadcast environments. While these protocols are useful in some applications, serializability may still be needed to guarantee the correctness of some other applications such as mobile stock trading where a buy/sell trade will be triggered to exploit the temporary pricing relationships among stocks. From the trader's perspective, the inability of database management system to maintain the serializability may have important financial consequences [3]. For instance, if the users who submitted multiple read-only transactions communicate and compare their query results, they may be confused [5].

2.2 Control Information

To guarantee serializability, additional control information has to be broadcast by the server to the mobile clients. The volume and the complexity of the control information affect the performance of the system. First of all, sending control information consumes communication bandwidth. Secondly, the volume of control information attributes to the length of broadcast cycle, which in turn has great impact on the response time of transactions at the mobile clients.

2.3   Early Restart of Transactions

As we have mentioned in Section 1, the basic optimistic approach may suffer from
the late restart of mobile transactions due to data conflicts with the committed
transactions at the server. It is desirable that transactions should be restarted
as soon as any data conflict is detected at the mobile clients. The concurrency
control protocol should enable the mobile clients to restart those transactions
that are destined to restart so that upstream communication bandwidth can be
saved.

3   Optimistic Approach for Broadcast Environments

3.1   Development of Thought

One of the design objectives of the protocol is that the new protocol should be
fully compatible to the conventional concurrency control protocol running at the
server. In this paper, we assume that the underlying concurrency control at the
server is the optimistic concurrency control. The key component in the optimistic
approach is the validation phase where a transaction's destiny is decided. Validation can be carried out in two ways: *backward validation* and *forward validation*
[6]. While in backward scheme, the validation process is done against committed transactions, in forward validation, validating of a transaction is carried out
against currently running transactions. Forward scheme generally detects and
resolves data conflicts earlier than backward validation, and hence it wastes less
resources and time.

In broadcast environments, there are different considerations. At the server,
there are transactions submitted for validation by different sources including the
mobile clients. It is not desirable to restart a validating transaction, in particular, a mobile transaction because of the high cost to restart a mobile transaction.
Therefore, forward validation is a better choice for the server because of the flexibility to choose the conflicting active transactions to be restarted. In addition,
only the write set of a transaction is required for forward validation, it allows
the read-only transactions to be validated locally and autonomously at the mobile clients. Given the serialization order determined by the server, the mobile
clients have to determine whether their mobile transactions can be committed
by detecting any data conflicts with the committed transactions. As a result, the
mobile clients have to carry out the backward validation process. Although it is
possible to carry out the backward validation at the end of a transaction execution, it causes delayed transaction restarts and more wasted resources at the
mobile clients. Therefore, partial backward validation is introduced at the beginning of every broadcast cycle. Any transaction that is found to read inconsistent
data will be restarted immediately.

3.2   Principles

In our proposed protocol, validation of transactions may perform at both the mobile clients and the server. The validation is based on the following principle to ensure serializability.

If a transaction T_i is serialized before transaction T_j, the following two conditions must be satisfied:

Condition 1: *No overwriting*
The writes of T_i should not overwrite the writes of T_j.

Condition 2: *No read dependency*
The writes of T_i should not affect the read phase of T_j.

Generally, Condition 1 is automatically ensured because I/O operations in the write phase are performed sequentially in critical section at the server. Thus, only Condition 2 will be considered and it is carried out in the following two ways based on where the validation is carried out.

3.3   Backward Validation at the Mobile Clients

At the mobile clients, all mobile transactions including read-only transactions and update transactions have to perform a partial validation at the beginning of a broadcast cycle before they access data objects in the cycle. The content of the control information will be described later. If the transaction fails the partial validation, it will be aborted and restarted. Otherwise, the next operation of the transaction can proceed. The partial validation process is carried out against committed transactions (at the server) in the last broadcast cycle. Data conflicts are detected by comparing the read set of the validating mobile transaction and the write set of committed transactions, since committed transactions precede the validating transaction in the serialization order. Such data conflicts are resolved to ensure Condition 2 by restarting the validating mobile transaction.

Let T_{pv} be the mobile transaction and $CD(C_i)$ be the set of data objects that have been committed (updated) in the last broadcast cycle C_i at the server. Let $CRS(T_{pv})$ denotes the current read set of transaction T_{pv}. In other words, $CRS(T_{pv})$ is the set of data objects that have been read by T_{pv} from previous broadcast cycles. Then the backward validation is described by the following procedure:

```
partial_validate(T_pv)
{
    if CD(C_i) ∩ CRS(T_pv) ≠ {} then
        abort(T_pv);
    else
    {
        record C_i;
        T_pv is allowed to continue;
    }
}
```

Note that $CD(C_i)$ is stored in the control information table that is broadcast at the beginning of every broadcast cycle.

3.4   Forward Validation at the Server

At the server, validation of a transaction is done against currently running transactions. Note that a validating transaction at the server may be a server transaction or a mobile transaction submitted by the mobile clients. This process is based on the assumption that the validating transaction is ahead of every concurrently running transaction still in read phase in serialization order. Thus, the write set of the validating transaction is used for data conflict detection to ensure Condition 2. For transactions already at the server during validation, the detection of data conflicts is carried out by comparing the write set of the validating transaction and the read set of active transactions. That is, if an active transaction, T_i, has read an object that has been concurrently written by the validating transaction, the value of the object used by T_i is not consistent and the conflicting active transaction, T_i, is aborted.

For an update transaction submitted by a mobile client, it has to perform a final backward validation before the forward validation. The final backward validation is necessary because there may be transactions committed since the last partial validation performed at the mobile client. Therefore, the broadcast cycle number of the last partial validation performed at the mobile client has to be sent to the server for final validation.

Let T_v be the validating transaction and T_a $(a = 1, 2, \ldots, n, a \neq v)$ be the conflicting transactions at the server in their read phase. Let C_k be the broadcast cycle number received along with T_v, if T_v is a mobile transaction. That is, T_v performs its last partial backward validation at the mobile client in the broadcast cycle C_k. Note that a mobile transaction may read or write data objects after partial validation and before being sent to the server for final validation. Let T_c $(c = 1, 2, \ldots, m)$ be the transactions committed at the server since C_k. Let $RS(T)$ and $WS(T)$ denote the read set and the write set of transaction T, respectively. Recall that $CD(C_i)$ is the set of data objects that are updated in the current broadcast cycle and is initialized at the beginning of every broadcast cycle. Then the forward validation is described by the following procedure:

```
validate(T_v)
{
    if T_v is a mobile transaction then
    {
        final_validate (T_v);
        if return fail then abort (T_v) and exit;
    }
    foreach T_a (a = 1, 2, . . . , n)
    {
        if RS(T_a) ∩ WS(T_v) ≠ {} then abort (T_a);
    }
```

```
        commit WS(T_v) to database;
        CD(C_i) = CD(C_i) ∪ WS(T_v);
}


final_validate (T_v)
{
    valid := true;
    foreach T_c (c = 1, 2, . . . , m)
    {
        if WS(T_c) ∩ RS(T_v) ≠ {} then valid := false;
        if not valid then exit loop;
    }
    if valid then return success
    else return fail;
}
```

If T_v is successfully validated, the write set of T_v is recorded in the control information table, which will be broadcast in the next broadcast cycle. The control information helps the mobile clients to perform local partial validation. In addition, the final validation results (commit or abort) of the mobile transactions are included in the control information table for acknowledgement to the mobile clients for further action.

3.5   The Roles of Server and Mobile Clients

The server performs the following functions:
1. During each broadcast cycle, it broadcasts the latest committed values of all data objects at the beginning of the broadcast cycle.
2. It uses the optimistic concurrency control with forward validation protocol described above to ensure serializability of all transactions submitted to the server.
3. It transmits the control information during each broadcast cycle that helps mobile clients perform the partial validation.

The mobile clients handle two types of transactions: read-only transactions and update transactions. Read, write, commit and abort operations performed by a transaction are handled as follows:
1. Read Operation: Before a read operation is performed on a data object broadcast during a cycle, the control information transmitted during that cycle is consulted to perform the partial validation. If the transaction fails the validation, the transaction is aborted. Otherwise, the read operation can proceed.
2. Write Operation: When a data object is written, the write is performed on a private workspace at the mobile client. No partial validation is performed.
3. Commit Operation: If it is a read-only transaction, then the commit operation succeeds. In case of an update transaction, the read set, write set, the values written, and the cycle number of the last partial validation performed at

the mobile client are sent to the server. The server performs the final validation to see whether the update transaction can be committed and sends the result to the mobile client.

4. Abort Operation: If it is a read-only transaction, then the abort operation does nothing. In case of an update transaction, then all the copies of the data objects written in the private workspace are discarded.

For a read-only transaction at the mobile client, it can be committed locally if it passes all the partial backward validation in the course of its execution. When a read-only transaction reads all the requested data objects and commits, it is serialized *after* all transactions committed before the beginning of the current broadcast cycle and is serialized *before* all transactions committed after the beginning of the current broadcast cycle. For an update transaction, it has to be sent to the server for final validation because the position of the update transaction in the serialization order is determined on the fly (at the validation point) at the server. That is, the update transaction is serialized *after* all transactions committed before its validation point and is serialized *before* all active transactions. After the final backward validation, forward validation is still required because those active transactions that have data conflicts with the validating transaction have to be identified.

3.6   Space Efficiency

The size of the control information required in our proposed protocol is relatively low compared to others [19], [16] in the literature. Let d be the size of a data object identifier, t be the maximum number of transactions committed during a broadcast cycle, and N be the maximum number of write operations per transactions at the server. There is a need for mobile transactions to know the broadcast cycle number in which the partial validation is performed. Then, we also broadcast the broadcast cycle number. Let c be the size of a broadcast cycle number. The total size of the control information is therefore $Ntd + c$.

4   Discussions and Future Work

The proposed protocol adapts from the conventional optimistic concurrency control protocols. While transactions at the server are processed using the forward validation, transactions at the mobile clients are processed using the partial backward validation. The development of the proposed protocol is motivated by a number of requirements and constraints in broadcast environments. In this section, we first discuss the features of our proposed protocol that are adaptable in broadcast environments. Then we describe some of our future research directions in optimizing the performance of our protocol.

4.1   Asymmetric Communications

The first most critical constraint in broadcast environments is the limited upstream bandwidth from the mobile clients to the server. Therefore, a desirable protocol should minimize the use of upstream communication channel. Locking-based protocols are not feasible because lock request for each data object from the mobile clients to the server will be too expensive. The transaction response time will be intolerably long due to the low bandwidth in wireless environments. Furthermore, the population size in broadcast environments is usually very large. Huge number of lock requests will overload the server. Therefore, optimistic concurrency control protocols are preferable. One straightforward approach to adapt optimistic concurrency control in broadcast environments is to submit all the required information of a transaction to the server for validation after the transaction finishes reading and pre-writing all the requested data objects at the mobile client. However, this method still suffers from a number of overheads. Large number of validation requests may overload the server and the server has to keep a long history record of committed transactions for validation. In our proposed protocol, read-only transactions can be validated and committed locally at the mobile clients without contacting the server. In view of the large proportion of read-only transactions in most data dissemination applications, the autonomy of mobile clients to process read-only transactions saves much processing burden of the server and upstream communication cost. For update transactions, part of the validation process is performed in advance so that only a final validation process is required at the server. It further alleviates part of the burden from the server.

4.2   Limited Capability of Mobile Computers

Another constraints are the limited capability and battery life of the mobile clients. It is of particular importance to reduce the amount of wasted processing resources in mobile computers. Therefore, it is desirable to abort a transaction as soon as any data inconsistency is found. In our proposed protocol, the partial backward validation helps to detect data inconsistency such that transactions can be aborted early instead of until the end of read phase. In addition, the validation also helps to identify those update transactions that are most likely to commit before transmitting them to the server for final validation such that the use battery power is more effective.

4.3   Dynamic Adjustment of Serialization Order

As we mentioned above, forward validation is based on an assumption that the validating transaction, if not restarted, always precedes concurrently running active transactions in the serialization order. This assumption is not necessary and can incur unnecessary transaction restarts. These restarts should be avoided [12]. In our future work, we will incorporate the mechanism of dynamic adjustment of serialization order in our new protocol in order to reduce transaction restarts.

4.4   Real-Time Conflict Resolution

In real-time mobile applications, transactions submitted by mobile clients may have priorities and data conflicts should be resolved in favor of higher priority transactions [14]. Since forward validation provides flexibility for conflict resolution that either the validating transaction or the conflicting active transactions may be chosen to restart, so it is possible to introduce priorities in broadcast environments, though it may be expensive to restart a mobile transaction than a server transaction. Some conflict resolution policies may be integrated into the partial backward validation process. For instance, a high priority update transaction may restart a low priority read-only transaction in a partial validation process, expecting that the update transaction will be committed at the server. However, careful selection of transactions to be aborted is important to the overall system performance because the cost to restart different transactions can be very different such as update transaction versus read-only transaction and mobile transaction versus server transaction.

5   Conclusions

In this paper, we have proposed a concurrency control protocol in broadcast environments. In this protocol, both read-only transactions and update transactions at the mobile clients are processed differently. At the server side, server transactions are processed by conventional optimistic concurrency control with forward validation. At the mobile clients, read-only transactions can be processed and committed locally without contacting the server. For update transactions, they are partially validated by consulting the control information broadcast in every cycle. The partial backward validation is performed against the committed transactions at the server in the last broadcast cycle. Upon the completion of the read phase, the update transactions will be sent to the server for final validation.

The proposed protocol is motivated by the special requirements and constraints of broadcast environments. The limited upstream bandwidth from the mobile clients to the server, the low bandwidth of wireless communication, the low capability and the short battery life of mobile computers are considered in the design of the protocol. Therefore, the protocol allows maximum autonomy between the mobile clients and the server. Read-only transactions can be processed locally without upstream communication. Most of the execution of update transactions is processed at the mobile clients. Only update transactions with maximum chance to commit will be sent to the server for final validation. The control information volume and the computation requirements at both the server and the mobile clients are low.

References

1. Acharya, S., Alonso, R., Franklin, M., and Zdonik, S., §Broadcast Disks: Data Management for Asymmetric Communication Environments,œ Proceedings of the ACM SIGMOD Conference, pp. 199-210, 1995.

2. Barbara, D., and Imielinski, T., §Sleepers and Workaholics: Caching Strategies in Mobile Environments,œ Proceedings of the 1994 ACM SIGMOD international conference on Management of data, pp. 1-12, 1994.

3. Bowen, T. F., Gopal, G., Herman, G., Hickey, T., Lee, K. C., Mansfield, W. H., Raitz, J., and Weinrib, A., §The Datacycle Architecture,œ Communications of the ACM, vol. 35, no. 12, pp. 71-81, 1992.

4. Dunham, M. H., Helal, A., Balakrishnan, S., §A mobile transaction model that captures both the data and movement behavior,œ Mobile Networks and Applications, vol. 2, pp. 149 § 162, 1997.

5. Garcia-Molina, H., and Wiederhold, G., §Read-Only Transactions in a Distributed Database,œ ACM Transactions on Database Systems, vol. 7, no. 2, pp. 209-234, 1982.

6. Haerder, T., §Observations on Optimistic Concurrency Control Schemes,œ Information Systems, vol. 9, no. 2, 1984.

7. Hayden, D., §The New Age of Wireless,œ Mobile Office, 1992.

8. Huang, Y., Sistla, P., and Wolfson, O., §Data Replication for Mobile Computers,œ Proceedings of the 1994 ACM SIGMOD international conference on Management of data, pp. 13-24, 1994.

9. Imielinski, T., and Badrinath, B. R., §Mobile Wireless Computing: Challenges in Data Management,œ Communications of the ACM, vol. 37, no. 10, pp. 18-28, 1994.

10. Kung, H. T., and Robinson, J. T., §On Optimistic Methods for Concurrency Control,œ ACM Transactions on Database Systems, vol. 6, no. 2, pp.213-226, 1981.

11. Lam, K. Y., Au, M. W., and Chan, E., §Broadcast of Consistent Data to Read-Only Transactions from Mobile Clients,œ Proceedings of Second IEEE Workshop on Mobile Computing Systems and Applications, 1999.

12. Lee, J., and Son, S. H., "Using Dynamic Adjustment of Serialization Order for Real- Time Database Systems," Proceedings of 14th IEEE Real-Time Systems Symposium, pp. 66-75, 1993.

13. Lee, V. C. S., Lam, K. Y., and Tsang, W. H., §Transaction Processing in Wireless Distributed Real-time Database Systems,œ Proceedings of the 10th Euromicro Workshop on Real Time Systems, pp. 214-220, Berlin, June, 1998.

14. Lee, V. C. S., Lam, K. Y., and Kao, B., §Priority Scheduling of Transactions in Distributed Real-time Databases,œ Real-time Systems, vol. 16, no. 1, pp. 31-62, 1999.

15. Pitoura, E., and Bhargava, B., §Building Information Systems for Mobile Environments,œ Proceedings of the third international conference on Information and knowledge management, pp. 371-378, 1994.

16. Pitoura, E., §Supporting Read-Only Transactions in Wireless Broadcasting,œ Proceedings of the DEXA98 International Workshop on Mobility in Databases and Distributed Systems, pp. 428-433, 1998.

17. Pitoura, E., and Samaras, G., Data Management for Mobile Computing, Kluwer Academic Publishers, 1998.

18. Shanmugasundaram, J., Nithrakasyap, A., Padhye, J., Sivasankaran, R., Xiong, M., and Ramamritham, K., §Transaction Processing in Broadcast Disk Environments,œ Advanced Transaction Models and Architectures, Jajodia, S., and Kerschberg, L., editors, Kluwer, Boston, pp. 321-338, 1997.

19. Shanmugasundaram, J., Nithrakashyap, A., Sivasankaran, R., and Ramamritham, K., §Efficient Concurrency Control for Broadcast Environments,œ (to appear in) ACM SIGMOD International Conference on Management of Data, 1999.

Session III: Ubiquitous Information Services

Modelling the WAP Transaction Service Using Coloured Petri Nets

Steven Gordon and Jonathan Billington

Cooperative Research Centre for Satellite Systems
University of South Australia
Mawson Lakes SA 5095, Australia
{sgordon,jb}@spri.levels.unisa.edu.au

Abstract. The Wireless Application Protocol (WAP) is an architecture designed to support the provision of wireless Internet services to mobile users with hand-held devices. The Wireless Transaction Protocol is a layer of WAP that provides a reliable request/response service suited for Web applications. In this paper Coloured Petri nets are used to model and generate the possible primitive sequences of the request/response Transaction Service. From the results we conclude that the service specification lacks an adequate description of what constitutes the end of a transaction. No other deficiencies were found in the Transaction Service.

1 Introduction

As wireless technologies advance, efficient access to Internet and advanced information services is becoming an important requirement from the perspective of mobile users. Currently, characteristics of wireless networks (e.g. low and varying bandwidths, drop-outs) and terminal devices (e.g. low power requirements, small displays, various input devices) limit the quality of these services for the mobile user. Therefore there is a need for existing protocols designed for use in the fixed network to be refined, and when necessary new protocols created that alleviate some of these limitations. The Wireless Application Protocol (WAP) [5] defines a set of protocols that aim to do this. In particular, the Wireless Transaction Protocol [6] provides a request/response service that is suited to using Web applications from hand-held devices such as mobile phones.

As with any new communication protocol, it is important to ensure the correctness of the Wireless Transaction Protocol. In this paper, Coloured Petri nets (CPNs) [8] are used to model the WAP Class 2 Transaction Service and generate it's language, the possible sequences of events between the users of the service and the service provider [2]. This can help identify any deficiencies in the current service specification. It is also the first step in the verification of the Wireless Transaction Protocol design. The use of formal methods is important because ensuring the correctness of a complex protocol is seldom possible via other design approaches. High-level Petri nets are a suitable formal method for the design of communication protocols because of their ability to express concurrency, non-determinism and system concepts at different levels of abstraction.

H.V. Leong et al. (Eds.), MDA'99, LNCS 1748, pp. 109–118, 1999.

They have been used to analyse various protocols [3]. CPNs are a popular form of high-level Petri nets that have extensive tool support [4, 9] for the design of systems, including protocols.

2   Wireless Transaction Protocol

The Wireless Application Protocol (WAP) architecture comprises 5 layers: transport, security, transaction, session and application. The Wireless Transaction Protocol (WTP) [6] provides 3 classes of service to the session layer: Class 0 – unreliable invoke message with no result message; Class 1 – reliable invoke message with no result message; and Class 2 – reliable invoke message with one reliable result message (this is the basic transaction service). This section describes the Class 2 Transaction Service in more detail.

Layer-to-layer communication is defined using a set of service primitives [7]. For the Transaction Service, the primitives occur between the WTP user and the WTP service provider . The sequences of primitives describe how WTP provides the Transaction Service. The WTP service primitives and the possible types are: TR-Invoke – req (request), ind (indication), res (response), cnf (confirm); TR-Result – req, ind, res, cnf; and TR-Abort – req, ind.

A transaction is started by a user issuing a TR-Invoke.req primitive. This user becomes the initiator of the transaction and the destination user becomes the responder. The responder must start with a TR-Invoke.ind. Table 1 shows the primitives that may be immediately followed by given primitives at the initiator and responder interfaces. For example, at the initiator a TR-Invoke.req can be followed by a TR-Invoke.cnf, TR-Result.ind, TR-Abort.req or TR-Abort.ind. There is no information given regarding the global behaviour of the service in the WAP specification [6].

Table 1. Primitive sequences for WAP Transaction Service

	TR-Invoke				TR-Result				TR-Abort	
	req	*ind*	*res*	*cnf*	*req*	*ind*	*res*	*cnf*	*req*	*ind*
TR-Invoke.req										
TR-Invoke.ind										
TR-Invoke.res		X								
TR-Invoke.cnf	X									
TR-Result.req		X*	X							
TR-Result.ind	X*			X						
TR-Result.res						X				
TR-Result.cnf					X					
TR-Abort.req	X	X	X	X	X	X	X			
TR-Abort.ind	X	X	X	X	X	X	X			

Note: the primitive in each column may be immediately followed by the primitives marked with an X. Those marked with an X* are not possible if the User Acknowledgement option is used.

Each of the primitives has several parameters. The TR-Invoke request and indication must include both source and destination addresses and port numbers. Other parameters are: User Data, Class Type, Exit Info, Handle, Ack Type and Abort Code. Of special significance is Ack Type. This parameter is used to turn on or off the User Acknowledgement feature. When on, an explicit acknowledgement of the invoke is necessary (i.e. TR-Invoke.res and TR-Invoke.cnf). Otherwise, the result may implicitly acknowledge the invoke.

3   Coloured Petri Nets

CPNs are a class of high-level nets that extend the features of basic Petri nets. The net consists of two types of nodes, *places* (ellipses) and *transitions* (rectangles), and directional arcs between nodes. An input arc goes from a place to a transition and an output arc vice versa. Places are typed by a *colour* set. For example, in Fig. 1 place Initiator has the colour set State. Places may be marked by a value from the colour set. These are known as *tokens*. The collection of tokens on a place is called it's *marking*, and the marking of the CPN comprises the markings of all places. Transitions and arcs can also have inscriptions which are expressions that, along with the tokens in places, determine whether a transition is *enabled*. A transition is enabled if sufficient tokens exist in each of its input places (as determined by the input arc inscriptions), and the transition inscription, or *guard*, (given in square brackets) evaluates to true.

In Fig. 1, TR-Invoke.req is enabled because NULL (the initial marking given to the right of the place) is in the only input place and it is also the arc inscription, and there is no guard shown for the transition (which implies the guard is always true). A subset of the enabled transitions can *occur*. The occurrence of a transition destroys the necessary tokens in the input places and creates new tokens in the output places, as given by the expressions on the arcs. The occurrence of TR-Invoke.req replaces NULL with INVOKE_WAIT in Initiator and creates Invoke in place InitToResp. When variables are used in arc inscriptions or guards, the values they are bound to on occurrence of a transition give, together with the transition name, a *binding element*.

The CPNs in this paper were edited, simulated, and partly analysed using Design/CPN [4]. Design/CPN allows the CPN to be drawn on separate pages to increase the readability of the net. One technique used to combine the different pages is known as *fusion places* which are copies of a place. Design/CPN may be used to interactively or automatically simulate the net, or to create an *occurrence graph*. An occurrence graph (OG) is a graph with nodes and arcs representing net markings and binding elements, respectively. A complete OG represents all possible states the CPN can reach. In Design/CPN, queries can be made on the OG to determine dynamic properties of the CPN (e.g. deadlocks, live-lock, bounds on places). An OG can also be viewed as a finite state automata (FSA) which, with appropriate analysis techniques can be used to give the language accepted by the CPN (where the binding elements are the alphabet), which in our case is the Transaction Service language.

4   Transaction Service CPN

4.1   Modelling Assumptions

The aim of modelling the Transaction Service is to generate the service language. That is, the possible sequences of service primitives between the user and provider are of major interest. With this in mind, several assumptions can be made to simplify the model.

The primitive parameters have no effect on the sequences of primitives. Therefore each primitive is modelled as a message (e.g. Invoke) which represents the primitive type and its parameters (e.g. Invoke{SrcAdr, DestAdr, ... }).

Only the general case when User Acknowledgement is off is modelled. As the possible primitives when User Ack is on is a subset of this general case (i.e. two of the primitive sequences are not allowed), the language will be a subset of the language generated from the model. This is straightforward to obtain.

The channel between initiator and responder can be separated into two directions of flow. The channel does not guarantee ordering of messages, hence each direction can be modelled as a single place. The modelling of the channel has also been done with the analysis techniques in mind. The OG calculated can be viewed as a FSA, which in turn can be minimised using a standard reduction technique [1]. Knowing this, it is possible to allow the reduction technique to handle functionality that would otherwise be necessary in the model. This is explained in detail in Sect. 5.1 after the model is presented.

4.2   CPN Model

The CPN model of the WAP Transaction Service has four separate pages (Fig. 1 to 4), representing an invocation, result, user abort and provider abort. Each page has the same structure:

- Two fusion places called Initiator and Responder representing the states of the initiator and responder, respectively. These places are typed by the colour set State:
  color State = with NULL | INVOKE_WAIT | INVOKE_READY | WAIT_USER | RESULT_WAIT | RESULT_READY | FINISHED | ABORTED;
- Two fusion places called InitToResp and RespToInit representing the communication channels from initiator to responder, and from responder to initiator, respectively. These places are typed by the colour set Message:
  color Message = with Invoke | Ack | Result | Abort;
- Transitions that represent the sending and receiving of the different primitive types by the user. Note that each transition has a boxed 'C' underneath. This indicates a *code segment* is used. Code segments are CPN ML code that are executed by Design/CPN when the associated transition occurs [10]. This allows, for example, auxiliary graphics to be drawn as the net is simulated. For the Transaction Service CPN code segments are used to draw message sequence charts (MSC) (e.g. Fig. 5(a)).

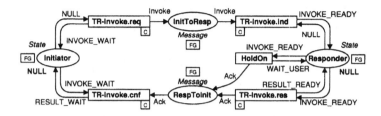

Fig. 1. TR-Invoke primitive sequence CPN

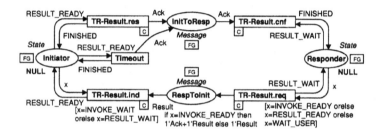

Fig. 2. TR-Result primitive sequence CPN

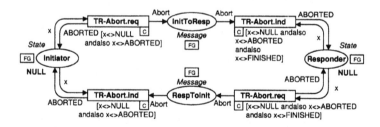

Fig. 3. User TR-Abort primitive sequence CPN

Fig. 4. Provider TR-Abort primitive sequence CPN

In addition, the Invoke and Result pages have transitions (HoldOn and Time-out, respectively) that indicate an interaction by the service provider which is not seen by the user (and therefore no primitive occurs). The occurrence of HoldOn, for example, indicates a timer has expired at the responder because the user is taking too long to generate the result. A message is sent to the initiator so that it will hold on until the responder user has generated the result. Finally, there is also a variable that can take any value from State:

var x:State;

Fig. 1 models the sequence of TR-Invoke primitives. The initial marking of both Initiator and Responder is NULL. In this marking the first and only transition that can occur is TR-Invoke.req. It follows that a possible transition occurrence sequence is the four TR-Invoke primitive types in order (i.e. request, indication, response then confirm). This would put the initiator into state RESULT_WAIT and the responder into RESULT_READY. From the TR-Result page (Fig. 2), again the TR-Result primitive transitions could occur in order. Both the initiator and responder would be in the FINISHED state. This sequence represents a successful transaction with explicit acknowledgement. The message sequence chart is shown in Fig. 5(a).

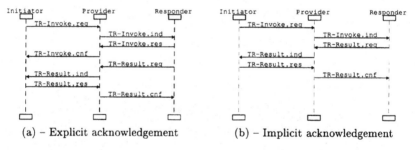

(a) – Explicit acknowledgement   (b) – Implicit acknowledgement

Fig. 5. MSC of service primitives for successful transaction

The Transaction Service does not require explicit acknowledgement of the TR-Invoke.req primitive by the responder (recall User Acknowledgement is assumed to be off). Instead by sending a TR-Result.req after receiving a TR-Invoke.ind, the responder can implicitly acknowledge the invocation. In Fig. 1 when the initiator is in the INVOKE_WAIT state and responder in INVOKE_READY both TR-Invoke.res and TR-Result.req are enabled. If TR-Result.req occurs (x is bound to INVOKE_READY) the Result message is sent to, and can be acknowledged by the initiator. The MSC for this sequence is shown in Fig. 5(b).

The previous two sequences of transitions were examples of successful transactions. However, as shown in Table 1, a TR-Abort.req or TR-Abort.ind from the initiator or responder can follow any primitive except themselves and, in the case of the responder, TR-Result.cnf (because the responder has successfully completed the transaction). Transaction aborts are modelled on two separate pages: one for user initiated abort (Fig. 3) and the other for provider initiated abort (Fig. 4). The aborts are symmetric – they can come from either initiator or responder. A separate page is used for the provider abort because the TR-Abort.ind does not require a message from either user. The channel places are shown to be consistent with the other pages – they are not necessary.

5 Analysis

The OG generated by Design/CPN for the Transaction Service CPN contains 85 nodes and 206 arcs. There were 28 different terminal states. The graph can be viewed as a FSA with the binding elements (essentially the service primitives) as the input language. Using a standard reduction technique [1], the minimised FSA gives a compact description of the possible sequences of primitives, or service language. This section explains how the minimisation of the OG can remove complexity from the model and presents the Transaction Service language.

5.1 Analysis Assumptions

The design of the model took into account the analysis techniques that would be applied (i.e. the FSA minimisation). In particular, it was expected that multiple terminal markings would be generated that were only differentiated by the markings of the places connecting the initiator to responder (InitToResp and RespToInit). Extra transitions could have been used in the model to remove all tokens from the these places once the initiator and responder had FINISHED or ABORTED. This would ensure only a single terminal marking was generated for these cases. However it was decided to let the FSA minimisation handle the extra terminal markings (it effectively merges all terminal markings into one) so the models could remain free of any "cleanup" transitions.

By treating the OG as a FSA, the introduction of halt states (states that indicate a possible end of a primitive sequence) was also possible. A halt state may or may not lead to other states. As well as all terminal markings being halt states, nodes were defined as halt states if they satisfied either of the following conditions:

- The marking of Initiator is ABORTED and the marking of InitToResp is Invoke. This represents the special case when a TR-Invoke.req is followed by a TR-Abort.req (by the user) or TR-Abort.ind (by the provider). This is a feasible halt state because the provider may not be able to notify the responder due to, for example, network failure. In this case the sequence of primitives is complete.
- The marking of Initiator is FINISHED and the marking of Responder is FINISHED or ABORTED. Although there is no mention of this in the service specification [6], we have assumed that when the initiator has acknowledged the result by issuing a TR-Result.res primitive (and the responder has finished or aborted) the transaction may be complete. However, aborts at the initiator are still possible.

5.2 Transaction Service Language

Fig. 6 shows the Transaction Service language obtained from the minimisation of the OG. There are 21 nodes and 74 arcs. For clarity, abbreviations of the service primitives are given for the arc labels. The first letter of each label represents the service primitive (*Invoke, Result, Abort*). The following three letters represent the primitive type (*request, indication, response, confirm*). In addition the initiator primitives are given in uppercase and the responder primitives in lowercase. Multiple arcs between two nodes are drawn as one with labels separated by commas. The combinations of TR-Abort.req and TR-Abort.ind from the initiator and responder are drawn as dashed and dotted lines, respectively.

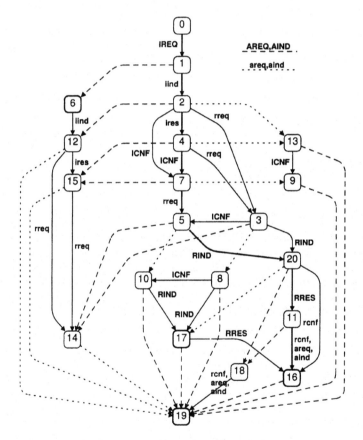

Fig. 6. Transaction Service language

There are four halt states in the language: nodes 6, 16, 17 and 19 (shown in bold). Node 6 represents the case when the initiator's TR-Invoke.req is immediately followed by an abort. Node 16 represents the case when the initiator has finished and the responder has also finished or aborted. Nodes 17 and 19 represent the cases when the transaction is aborted.

Further analysis reveals there are 450 possible sequences of primitives. The shortest sequences are 2 primitives (TR-Invoke.req followed by TR-Abort.req or TR-Abort.ind at the initiator) and the longest sequences are 9 primitives (e.g. a successful transaction (shown as bold arcs – this corresponds to the sequence shown in Fig. 5(a)) followed by a TR-Abort.req from the initiator).

The primitives between the following nodes are not possible when User Acknowledgement is turned on: (2,3), (2,7), (2, 13), (12,14), (8, 17), (3,20). In addition, the primitives that were between 2 and 13 are now between 2 and 9.

From the service language it is unclear why the initiator would issue a TR-Abort.req (e.g. node 16 to node 19 in Fig. 6) after it had acknowledged the result with a TR-Result.res primitive (e.g. node 20 to node 11). However, an examination of the protocol specification provides an explanation. Transaction information is saved after the initiator has sent the last acknowledgement in case retransmissions are necessary. By issuing a TR-Abort.req after a TR-Result.res, the transaction state information is released. Otherwise, a timeout will release the information.

6   Conclusions

We have described, modelled and analysed the WAP Class 2 Transaction Service in a first step towards verifying the Wireless Transaction Protocol. WTP utilises the datagram service (Transport layer) in the WAP architecture and provides a reliable request/response service to the upper layers.

Coloured Petri nets were used to model the Transaction Service and generate the occurrence graph. The knowledge of the analysis techniques used allowed several assumptions to be made that simplified the model. Halt states were introduced and the OG was treated as a finite state automata and reduced to obtain the service language.

The Transaction Service language generated provides a complete set of service primitive sequences, when taking both ends of the transaction into account (i.e. both initiator and responder). This global behaviour is not described in the WAP specification. From the modelling and analysis, two questions not fully answered in the service specification arose:

1. What constitutes the end of a transaction?
2. Why is a TR-Abort.req primitive possible after a TR-Result.res from the initiator user?

Answers were obtained from examining the operation of the protocol in more detail. A transaction may be considered complete if either:

1. both initiator and responder have aborted,
2. a TR-Invoke.req at the initiator is followed immediately by an abort, and the provider hasn't notified the responder (e.g. due to network failure),

3. the initiator has acknowledged the result and the responder has either received the acknowledgement or aborted.

In the final case, it is still possible for the initiator to issue a TR-Abort.req to clear transaction state information. The need to understand the protocol operation is a shortcoming of the specification – the service should be described independently of the protocol. No other deficiencies have been found in the service specification.

The Transaction Service language can be used as a basis for verifying that the Wireless Transaction Protocol conforms to the service specification. The next step to achieve this is to model the operation of the protocol in detail. This work is in progress. An incremental approach is being used so different features can be modelled and analysed. The desired results are to generate an OG from which properties of the protocol can be derived (e.g. presence of deadlocks). Then the protocol language can be generated and compared to the service language.

Acknowledgements

This work was carried out with financial support from the Commonwealth of Australia through the Cooperative Research Centres Program.

References

1. W. A. Barret and J. D. Couch. *Compiler Construction: Theory and Practice.* Science Research Associates, 1979.
2. J. Billington. Abstract specification of the ISO Transport service definition using labelled Numerical Petri nets. In H. Rudin and C. H. West, editors, *Protocol Specification, Testing, and Verification, III*, pages 173–185. Elsevier Science Publishers, Amsterdam, New York, Oxford, 1983.
3. J. Billington, M. Diaz, and G. Rozenberg, editors. *Application of Petri Nets to Communication Networks: Advances in Petri Nets.* LNCS 1605. Springer-Verlag, Berlin Heidelberg New York, 1999.
4. Meta Software. *Design/CPN Reference Manual, Version 2.0.* 1993.
5. WAP Forum. Wireless application protocol architecture specification. Available via: http://www.wapforum.org/, Apr. 1998.
6. WAP Forum. Wireless application protocol wireless transaction protocol specification. Available via: http://www.wapforum.org/, Apr. 1998.
7. ISO/IEC. *Information Technology - Open Systems Interconnection - Basic Reference Model - Conventions for the Definition of OSI Services.* 10731. 1994.
8. K. Jensen. *Coloured Petri Nets. Basic Concepts, Analysis Methods and Practical Use*, Vol. 1-3. Springer-Verlag, Berlin, 1997.
9. K. Jensen, S. Christensen, and L. M. Kristensen. *Design/CPN Occurrence Graph Manual, Version 3.0.* Department of Computer Science, Aarhus University, Aarhus, Denmark, 1996.
10. L. M. Kristensen, S. Christensen, and K. Jensen. The practitioner's guide to Coloured Petri nets. *Int J Software Tools for Technology Transfer*, 2(2):98–132, 1998.

Enabling Ubiquitous Database Access with XML

Hui Lei[1], Kang-Woo Lee[2], Marion Blount[1] and Carl Tait[1]

[1] IBM Thomas J. Watson Research Center
Route 134
Yorktown Heights, NY 10598, USA
{hlei, mlblount, cdtait}@us.ibm.com
[2] Department of Computer Engineering
Seoul National University
Seoul 151-742, Korea
kwlee@oopsla.snu.ac.kr

Abstract. This paper describes the design and implementation of DataX, middleware for enabling remote database access from heterogeneous thin clients. Unlike existing commercial offerings that require a standalone database on the client side, DataX partially replicates the server database in the form of XML, using a weak consistency criterion. It also adapts data replication to device characteristics and user preferences. It employs a per-device renderer to present data in a form layout, making the data access semantics separate from user interaction details and independent of the device type. It allows for rapid development of end-to-end solutions and application portability across multiple client and server platforms.

1 Introduction

Many industries employ mobile workers who perform a significant part of their work away from the office. Examples of such mobile workers are insurance agents, visiting nurses and field service technicians. As the population of the mobile workforce rapidly grows, it is becoming increasingly important for corporations to extend enterprise information and applications to their mobile workers. That will enable the latter to be more productive while on the road, to respond to customers faster and to improve customer satisfaction.

Until recently, notebook computers and modems have been the main computing and communication resources for mobile workers. Thin clients, ranging from palmtop computers to personal digital assistants (PDAs) to smart cellular phones, have only been used for personal information management (PIM) applications. Due to their form factor, low cost of ownership, and ease of use, thin clients are quickly emerging as a new wave of mobile computing devices. Many organizations are beginning to exploit these thin clients to meet the specific needs of their mobile employees.

Since mobile computers often operate for extended periods over low bandwidth links or without connectivity at all, disconnected operation is a key enabling technique for mobile data access. The basic mechanism for disconnected

H.V. Leong et al. (Eds.), MDA'99, LNCS 1748, pp. 119–134, 1999.

operation is well understood. Before disconnection, a subset of server data needs to be replicated on the local device. While disconnected, the user works with the local copy. Changes made to the local replica are reintegrated into the server's copy when a connection is established again. Any implementation of the above scheme, however, needs to address three issues:

Hoarding: In general, the system alone is not able to decide which subsets of the backend data should be locally replicated. This is because future user behavior is not perfectly predictable [20, 30]. External assistance is often solicited, either from end users or from system administrators. To minimize the burden placed upon users/administrators, it is desirable to limit their participation to a high semantic level by providing some sort of data encapsulation mechanism [22].

Adaptation: Data as stored in the server database sometimes may need to be transformed to make it appropriate in a mobile environment. Reasons for such data adaptation may be that the destination device is not capable of hosting the data as it is, or that the user is willing to accept data of lower quality in order to reduce consumption of precious client resources [9, 25].

Synchronization: Data consistency needs to be preserved between the partial replica on the client and the full replica on the server. A replication strategy should be adopted that controls what kind of data consistency is provided and how divergent copies are integrated [4].

A thin-client-based mobile solution imposes further challenges. The variety of thin client devices is impressive; these devices vary in hardware capacity and software functionality. Their relatively low cost is likely to cause the pervasive deployment of such devices. A company often finds it necessary to support multiple device types, for reasons such as users' choice, continuous availability of new and better devices, reorganizations and mergers. This in turn translates into two requirements: the server side of the solution has to be truly open and interoperable, and porting client-side applications to new device types must be easy.

The DataX project developed at the IBM T. J. Watson Research Center is middleware that provides for disconnected database access from heterogeneous thin clients. Currently it supports only data access scenarios where strict transactional properties are not required. Many enterprise database applications such as data viewers and data collection fall into this category. DataX is based on a three-tiered client/proxy/server architecture. It introduces an abstraction mechanism, called *folders*, to facilitate data hoarding. It employs a rule-based targeting engine to adapt data according to client characteristics and user preferences. It represents client replicas in the form of platform-neutral XML and makes use of a per-device renderer to present data and control user interaction. Finally it adopts a weak-consistency replication strategy and resolves update conflicts based on data semantics.

In an earlier position paper [23], we described the overall DataX architecture and discussed the principal considerations behind our design. Since then we have implemented a prototype of DataX. In this paper, we shall present our refined design and the details of implementation. The rest of the paper is organized as follows. Section 2 examines the major system components of DataX. Section 3 details the use of XML in defining and representing folders, the basic unit of client-side replication. Section 4 discusses related work in various areas. Our conclusions are presented in Section 5.

2   System Components

Figure 1 shows the basic DataX system architecture. The system consists of three tiers: client, proxy and server, where the proxy is placed at the fringe of the fixed network and provides a single point of connection for heterogeneous types of client devices. Such a tiered model is common in mobile systems because it confers three advantages. First, it alleviates the impact of client resource poverty by migrating computational and administrative tasks from the client to the proxy. Second, it off-loads mobile-specific responsibilities from the server to the proxy so that the server may concentrate on servicing the general user community. Third, it accommodates weak connectivity and disconnected operation via asynchronous processing and message queuing.

In DataX, data flows along the following path. The subsetting component on the proxy extracts necessary data from the server database and pipelines it to the targeting component for adaptation. The output of adaptation is stored as a client mirror, which takes the form of an XML document. The proxy-side peer synchronizer compares the old and current mirrors for the client and sends the differences to the client device. The partial data replica on the client is also represented in XML. A generic rendering component allows the user to view the data on the device and perform updates. At the user's discretion, the client-side peer synchronizer transmits the client updates to the proxy, which are then incorporated into the server by the reintegration component.

The proxy accesses the server database through the standard JDBC interface so that the system is not bound to any specific server platform. The peer synchronizers communicate via a proprietary protocol [24]. Each message exchanged is an XML document specifying the data changes on one side. A detailed description of the peer synchronizers is beyond the scope of this paper. In the remainder of this section, we take a closer look at other key components of the system: subsetting, targeting, reintegration and rendering.

2.1   Subsetting

The subsetting component is concerned with what portion of the server database should be hoarded, a process that requires external input. In order to allow end users or system administrators to make selections at a high semantic level, we institute an encapsulation mechanism called *folders*. Folders present a high-level view of data without exposing the details of data organization. A folder is a parametric logical collection of data. A folder not only describes a data subset, but also the semantics of data access.

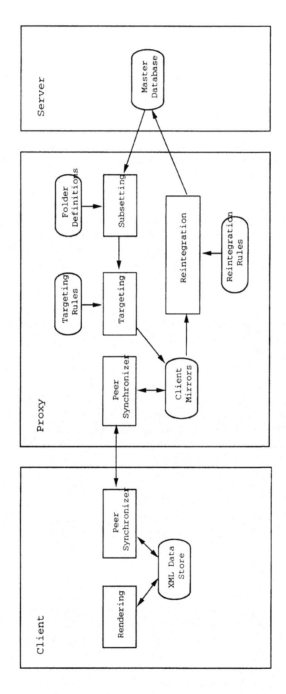

Fig. 1. DataX Architecture

Instead of developing a full-fledged application, a programmer simply writes a folder definition. A folder definition consists of four sections: parameters, data, operations, and constraints. The parameters section specifies the folder parameters and their optional default values. The data section qualifies the database records that belong to the folder by defining a view to the database, where the selection criterion is a predicate on the folder parameters. The operations section enumerates the data access operations a user is allowed to perform. At the record level, it specifies whether records are read-only or updatable and, if updatable, the operations permitted (insert, delete and/or modify). At the field level, it specifies whether each field is updatable and whether records can be sorted on the field. Finally the constraints section specifies the integrity constraints for the folder. These are the constraints that are to be verified whenever records are updated on the client. They are separate from the constraints that are part of the backend database definition and that are enforced by the database server. We shall revisit the definition of a folder in Section 3.

We envision two kinds of application scenarios. In some cases, data subsets have to be dynamically selected before each disconnected session, either by the mobile user at the client or by the system administrator at the proxy. The user or the administrator selects from a list of folder names. He may accept the default parameter values or choose to provide his own. Other situations are more straightforward, where the hoard selection is static. The client device can simply be configured with the folders it needs to replicate and the parameter values that should be used. The user is spared from intervention on a per-disconnection basis.

The subsetting component is composed of a compiler and an extractor. The compiler takes a folder definition as input, verifies its validity and ensures that the definition is compatible with the schema of the server database. The compiler also translates the definition to an internal format. The extractor is executed every time a data subset needs to be generated, with a folder name and the parameter values as its input. The extractor uses information gathered by the compiler and retrieves relevant records from the server database. These records, combined with meta-information such as permissible operations and constraints, are then passed to the targeting component.

2.2   Targeting

A folder determines the capability of its users: what data is accessible and how it can be accessed. Prior to transmission to the client device, however, records that comprise a folder instance are subject to data transformation by the targeting component to fit the device's characteristics and/or the user's preferences.

Roughly speaking, data transformation falls into one of two categories:

Filtering: Selective transmission of unmodified data.
Transcoding: Changing the fidelity and/or modality of a particular piece of data.

An example of filtering is transferring only some of the columns of a relational table. Truncating a text field beyond a certain size and changing a rich text into plain ASCII are classed as fidelity transcoding; converting text into speech is a form of modality transcoding. For semantic reasons, transcoding is only performed on read-only data.

The behavior of the targeting component is controlled by a set of rules. Each rule specifies an action and a condition under which the action may be triggered. The condition is a conjunction of one or more predicates, each of which tests one of the following:

- Device type
- Folder name
- Field name
- Data type
- User identification

Each targeting action is a data transformation function registered with DataX.

For common device types, DataX specifies default targeting policies. Application developers can introduce rules that are specific to an application (folder). Further, individual users may define rules that are tailored to their own needs. User-specific rules may or may not be application-specific. Here are some examples of targeting rules.

1. A general rule: *for a WorkPad, always convert rich text to ASCII.*
   DevType=WorkPad ∧ **Data**=RichText → Rich2Ascii
2. An application-specific rule: *When transferring customer records, truncate notes beyond 100 characters long.*
   Folder="customers" ∧ **Field**="notes" → Truncate(100)
3. A user-specific rule: *when the customers folder is accessed by user Smith on a WorkPad, filter out information on past customer visits (stored in the activities field).*
   DevType=WorkPad ∧ **Folder**="customers" ∧ **User**=Smith →
   Exclude("activities")

The targeting process for any folder instance consists of two phases. In Phase 1, an applicable filtering rule, if found, is triggered. In Phase 2, for each data field, an applicable transcoding rule, if found, is triggered. In each phase, a rule is first searched for among user-specific rules, then among rules that are application-specific but not user-specific, and finally among general rules.

2.3   Reintegration

The role of the reintegration component is to modify the server's database to reflect any changes made on the mobile client. A change can be an insertion, an update, or a deletion of a database record. Both the initial image and the final image of changed records are available to the reintegration component. The initial image of a record refers to its state at the time of replication, which can be found in the client mirror that is stored on the proxy. Initial images are required to detect update conflicts, as explained below. The final image of a changed record reflects its current state on the client and is communicated via the peer synchronizers.

It is widely recognized that a strong-consistency replication strategy that ensures one-copy serializability is not suitable for mobile computing. Enforcing strong consistency either incurs very high logging and repair costs, as in an optimistic strategy, or severely limits data availability, as in a pessimistic strategy [4]. Hence DataX adopts a weak consistency replication strategy, which only ensures that replicas eventually reach identical state or become mutually consistent. Such a strategy is required to support disconnected operation because it allows for high availability, good scalability and simplicity of the system.

If a changed record does not cause update conflicts, the change is simply replayed on the server, subject to any integrity constraints enforced by the server database. If an update conflict occurs, the reintegration will try to resolve the conflict based on data semantics. An update conflict can be either an update/update conflict or an update/delete conflict. However, the two are treated conceptually in the same way by associating a bit field with each record to indicate whether the record has been deleted. An update conflict occurs if and only if a record is changed on the client and the record image in the server database is not identical to the initial image as found in the client mirror.

Two types of semantic information are needed to resolve update conflicts [10]. The first type is concerned with computed fields, $i.e.$, values in a database that are computable from the rest of the values. All computed fields should be identified along with a formula for computing them. Obviously, there can be no circular definitions. Computed fields are evaluated at the end of the conflict resolution process, based on the reintegrated value of non-computed fields. The second type of information is reintegration rules. There should be an reintegration rule for each non-computed field, describing how to handle conflicting updates to the field. Assume that a non-computed field F had initial value f and that concurrent updates have resulted in value f_c on the client and f_s on the server. The reintegration rule can specify any one of the following as a resolution policy:

max (min): The maximum (minimum) of f_c and f_s becomes the reintegrated value of F.

client (server): The reintegrated value is provided by f_c (f_s).

constant k**:** If f_c and f_s are identical, then store that value in F. Otherwise, use constant k. This is useful for storing a null value if desired.

operational: The reintegrated value is set to $f_c + f_s - f$ ($i.e.$, deltas on both sides are applied to the original value).

reject: If f_c and f_s differ, reject the changes made on the client and alert the user.

program p**:** A program p is executed, whose parameters are f_c, f_s and f.

Although the last resolution option (program escape) represents the most general case, special options are explicitly provided as part of the DataX interface to save the programming effort of application developers.

2.4 Rendering

The rendering component on the client device retrieves data from the XML data store and presents it in a manner appropriate for the device. Recall that each XML document represents a folder instance, which describes a data subset as well as the semantic information on data access. The rendering component consists of a parser and a browser. While the former parses an XML document into a Document Object Model (DOM) [31] tree, the latter displays the data according to the characteristics of the device and controls the user interaction.

The browser is device-specific, but independent of which folder it manipulates. There is one generic browser for each category of client device. For a table that belongs to a folder, the browser displays a list of its records. The list may be sorted on various fields, as specified in the folder definition. The user can browse through the list or search for specific records. He can also expand individual records to get a more detailed view or to perform updates. All updates are subjects to the specified integrity constraints. If a folder contains multiple tables, the browser also allows the user to navigate through them.

The per-device rendering component frees the application developers from writing separate programs for each combination of applications and device types. Further, it significantly simplifies development effort. Developers can concentrate on the semantics of data access without having to worry about the intricacies of user interface design. Thus, development time is reduced and quality is enhanced.

3 Folders and XML

The folder is an abstraction for a parametric data subset and the underlying data model. Both the folder definition and the folder instance are coded in XML. Three facts make defining a folder in XML easy. First, XML is a descriptive markup language. It is simpler to program than a procedural scripting language or programming language. Second, when defining a folder, application developers make no distinctions regarding the particular device types to be supported or individual users' customization requirements. Third, developers can concentrate on the semantics of data access, leaving user interface issues to the rendering component.

The XML representation of folder instances brings the following benefits. Since the representation format is platform-neutral, the proxy-side computation is substantially simplified when preparing and reintegrating client replicas. In addition, application deployment does not rely on the availability of standalone database systems on the client, and system administrators are spared from maintaining such products. Finally, the self-describing nature of an XML document makes it possible to adapt the client replica according to device attributes and user's preferences.

3.1 Folder Definition

A folder definition may contain the following sections: parameters, data, operations and constraints. The data section is required, and the others optional. A

```
<!DOCTYPE folderDef SYSTEM "folderDef.dtd">
<folderDef name="visits">
   <parameters>
      <parameter name=":nurse_id">
   </parameters>
   <data database="pms">
      <table name="patients" schema="pms">
         <columns names="ssn name age address physician"/>
         <condition> "visiting_nurse=:nurse_id" </condition>
         <table name="physicians" schema="pms">
            <columns names="id name phone office"/>
            <condition> "physician=id" </condition>
         </table>
      </table>
   </data>
   <operations>
      <insert/>
      <modify columns="age address"/>
   </operations>
   <constraints>
      <range column="age" ge="0" le="150"/>
   </constraints>
</folderDef>
```

Fig. 2. Example of a Folder Definition

sample folder definition is given in Figure 2, which defines a folder called visits. A visits folder includes information on all the patients a visiting nurse covers. (The example is admittedly contrived since we are trying to show as many features as possible.)

The folder has one parameter :nurse_id, for which there is no default value. The data section is associated with an attribute **database** that identifies the name of the master database. Each **table** element specifies a base table in the master database. Both attributes **name** and **schema** are required to identify a base table. The visits folder is derived from two base tables **patients** and **physicians**. The condition element in table **patients** specifies a selection predicate, which is an equality test on the column visiting_nurse and the folder parameter :nurse_id. The condition element in **physicians**, on the other hand, specifies a join condition between the two base tables.

Note that the **data** section explicitly specifies the interrelationships between the base tables. This is necessary for the rendering component to control navigation between the tables. In our case, for instance, the renderer will start with records of **patients** and allow the user to follow links to view information on a patient's physician. A folder data set is essentially a materialized view defined on master database tables. In general, updating a view derived from multiple tables is not allowed due to the potential ambiguity [5]. We avoid this problem by imposing the restriction that only the outermost table, referred to as the *primary table*, may be updated.

Read access to data is always granted. The operations section in addition declares that records in patients can be inserted and modified and that modifications may be made to the columns age and address only. The constraints section specifies that the updated age value has to fall between 0 and 150.

Information on folder parameters is used in the folder selection process. Information in the data section guides the data extraction by the subsetting component. The operations and constraints sections are duplicated in each generated folder instance to help the rendering component control the user interaction.

3.2   Folder Instances

While folder definitions are coded by programmers, folder instances are produced automatically by the data extractor of the subsetting component. A folder instance contains both meta-information and data records. Meta-information describes permissible operations and integrity constraints, as well as table keys and columns. The data records are those in the master database that satisfy the conditions set forth in the folder definition. Figure 3 shows a folder instance that could be generated from the folder definition in Figure 2.

The operations and constraints sections are identical to those in the folder definition. The dataSet section is composed of a series of table elements, each corresponding to a base table and the first one being the primary table. A table construct contains metadata such as the primary key and column types that are taken from the table schema, and the data records that have been selected. Each selected record is represented in a row element, where all non-null column values are tagged by the column name. The row element also carries an attribute id, which is unique in the scope of the entire folder instance document and is generated by concatenating the table name and the primary key value.

Correlations between records from potentially different tables are expressed using the hypertext linking capability proposed by XML Linking Language (XLL) [32]. A source record contains link elements that point to target records. Each link element specifies the primary key value of the target in its text, and the row id in the href attribute. For each source record, link elements that describe target records from the same table are grouped within a linkGroup element. In terms of XLL, a link is a locator element and a linkGroup is an extended element. The portion of DTD that declares these constructs is listed in Figure 4. The use of extended/locator links allows arbitrary record correlations (1-1, 1-N, N-1) to be expressed.

```
<!DOCTYPE folderInstance SYSTEM "folderInstance.dtd">
<folderInstance name="visits">
   <operations>
      <insert/>
      <modify columns="age address"/>
   </operations>
   <constraints>
      <range column="age" ge="0" le="150"/>
   </constraints>
   <dataSet>
      <table name="patients">
         <primaryKey columns="ssn"/>
         <columns>
            <column name="ssn" type="VARCHAR"/>
            <column name="name" type="VARCHAR"/>
            <column name="age" type="INTEGER"/>
         </columns>
         <records>
            <row id="patients:210-95-7638">
               <ssn>210-95-7638</ssn>
               <name>Bill Smith</name>
               <age>35</age>
               ...
               <linkGroup>"physicians"
                  <link href="#physicians:99">99</link>
               </linkGroup>
            </row>
            <row> ... </row>
            ...
         </records>
      </table>
      <table name="physicians">
         <primaryKey columns="id"/>
         <columns> ... </columns>
         <records>
            <row id="physicians:99">
               <id>99</id>
               <name>Sidney Winter</name>
               ...
            </row>
            ...
         </records>
      </table>
   </dataSet>
</folderInstance>
```

Fig. 3. Example of a Folder Instance

```
<!ELEMENT linkGroup (#PCDATA, link+)>
<!ATTLIST linkGroup xml-link CDATA #FIXED "extended">
<!ELEMENT link (#PCDATA)>
<!ATTLIST link xml-link CDATA #FIXED "locator"
          href CDATA #REQUIRED>
```

Fig. 4. DTD for links

4 Related Work

Although many aspects of DataX have appeared in previous systems of various contexts, we view our contribution as the generalization and engineering of existing techniques into a uniform architecture that effectively supports database access from heterogeneous types of mobile devices. In this section, we discuss work in the areas our system spans.

Several methods have been proposed to enable mobile information access. They all rely on optimistic replication control to provide high data availability and performance. The notion of disconnected operation was successfully demonstrated in Coda [20], a client/server-based distributed file system. Ficus [13] is another file system that effectively supports primarily disconnected mobile computers, but adopts a peer-to-peer model. Both Coda and Ficus use version vectors to detect conflicts and a rule-based approach to select application-specific conflict resolvers. In contrast, Bayou [6] utilizes application-provided dependency checks and merge procedures to handle conflict detection and per-write conflict resolution respectively. For document accesses, Lotus Notes [18] allows conflicts to be automatically merged at both the document level and field level. The Rover toolkit [17] provides two mechanisms, relocatable dynamic objects and queued remote procedure calls, for building mobile-aware applications and writing proxies to enable legacy applications for mobility. IBM's eNetwork Web Express [8] combines caching and request queuing to support disconnected and asynchronous Web browsing. The need for dynamic adaptation in mobile information access has been identified by a number of projects. In particular, Odyssey [25] provides system support to enable mobile-aware applications to use data fidelity levels to control resource utilization.

Many researchers have looked into database system issues that arise in a mobile environment. Imielinski and Badrinath suggested solutions to the problems of location updates and location-dependent query processing [15]. An analysis of various static and dynamic data replication methods were presented in [14]. Barbara and Imielinski proposed an approach to invalidating the secondary replication on mobile clients [2], in which the server periodically broadcasts a report that reflects the changing database state. Krishnakumar and Jain discussed an architecture and protocols for mobile sales applications [21]. All these projects assume that the mobile clients are connected, via a wireless channel, most of the time. For disconnected operation, Badrinath and Phatak proposed clustering of data on the server based on the concept of hoard key [1]. Gray *et al.* presented a two-tier replication scheme to allow disconnected applications to request tentative transactions that are later applied to a master copy [11]. The conflict resolution mechanism in our system is based on the Data-Patch tool described in [10]. Major database vendors have just begun to offer standalone databases for hosting a portion of the backend database on mobile computers [26, 29]. However, these systems are not portable across system platforms, and they do not provide a uniform framework for adapting the data as we do.

As far as accommodating client variations is concerned, DataX has an overlap with DIANA [19] and GloMop [9]. DIANA advocates a form-based approach to structuring mobile applications. Applications in DIANA only need to be defined in terms of forms inputs requested from users and the application's responses to those users. Controlling user interaction in a device-specific manner is handled by a generic User Interface Logic component, thus de-coupling the user interface logic from the processing logic of applications. Disconnected operation is supported via hoarded forms and application surrogates that emulate the behavior of the applications. DataX focuses on database accesses, for which a form-based interface is most natural. There is no need for separate application surrogates in DataX, as data access semantics is declaratively specified in the folder instance and interpreted by the rendering component. DataX takes an intent-based approach to UI definition where all the interaction is encoded in the data model and the presentation is almost entirely determined by the renderer. DIANA is more prescriptive in interaction style and still leaves it to the renderer to choose the most appropriate representation for each interaction item. It would be possible to devise a UI definition language that offers the UI writer a spectrum of options in the amount of specificity he cares to exercise [28].

GloMop is a proxy architecture for enabling effective access to Internet content for a wide range of clients. The proxy retrieves documents on the client's behalf and performs transcoding on the fly to suit network characteristics and device capability. GloMop's emphasis is the read access of multimedia data (images and videos). Although its framework supports transcoding, DataX is mainly concerned with enterprise data, which is mostly text. DataX prepares those data for mobile access via subsetting and filtering. Since updates are part of life, DataX also provides a mechanism to integrate divergent copies. Finally, data adaptation policies in GloMop are specified by client applications through a special API. For legacy applications, a client-side agent is required to communicate with the proxy. In comparison, a rule-based interface for defining adaptation policies, like the one used in DataX, not only makes it easy to support legacy applications, but also allows for customized adaptation.

5 Conclusions

DataX is a proxy-based architecture that extends the reach of enterprise data to the users of thin mobile clients.It addresses the issues of data hoarding, adaptation, synchronization and client heterogeneity in the context of ubiquitous database access. It supports disconnected operation by replicating necessary data on the client device as instances of folders. A folder describes a data subset as well the data model that underpins the interactions with a user, leaving detailed presentation decisions to a device-specific renderer. Data replication is adapted according to device attributes and user preferences. Divergent client replicas are reintegrated into the server through an automated process.

XML plays a key role in DataX. It allows application behavior to be coded declaratively and transported to the client as directives to the per-device renderer, making it possible to provide application portability across heterogeneous client devices and to introduce new device types. It allows the client-side replica to be represented in a platform-neutral format, simplifying the computational and administrative tasks on the proxy. Further, the self-describing nature of an XML document allows targeting policies to be made on a per device/user basis, which leads to better utilization of client resources.

It is easy to deploy end-to-end mobile solutions with DataX, as application developers can concentrate on the semantics of data access, without having to worry about the infrastructure technologies, specific device characteristics, or user interface issues. This can be a major competitive advantage for organizations in today's world of pervasive computing and e-business systems.

References

1. B. R. Badrinath and S. Phatak. Database Server Organization for Handling Mobile Clients. Technical Report DCS-TR-324, Department of Computer Science, Rutgers University, 1997.
2. D. Barbara-Milla and T. Imielinski. Sleepers and Workaholics: Caching Strategies in Mobile Environments. *ACM SIGMOD Record*, 23(2), May 1994.
3. M. Butrico, H. Chang, A. Cocchi, N. Cohen, D. Shea, and S. Smith. Gold Rush: Mobile Transaction Middleware with Java-Object Replication. In *Proceedings of the 3rd Conference on Object-Oriented Technologies and Systems*, Portland, Oregon, June 1997.
4. S. B. Davidson, H. Garcia-Molina, and D. Skeen. Consistency in Partitioned Networks. *Computing Surveys*, 17(3):341–370, September 1985.
5. U. Dayal and P. Bernstein. On the Updatability of Relational Views. In *Proceedings of Fourth International Conference on Very Large Data Bases*, pages 368–377, September 1978.
6. A. Demers, K. Petersen, M. Spreitzer, D. Terry, M. Theimer, and B. Welch. The Bayou Architecture: Support for Data Sharing Among Mobile Users. In *Proceedings of the IEEE Workshop on Mobile Computing Systems and Applications*, Santa Cruz, CA, December 1994.
7. A. Elmagarmid, J. Jing, and T. Furukawa. Wireless Client/Server Computing for Personal Information Services and Applications. *ACM SIGMOD Record*, 24(4):16–21, 1995.

8. R. Floyd, B. Housel, and C. Tait. Mobile Web Access Using eNetwork Web Express. *IEEE Personal Communications: Special Issue on Mobile Access to Web Resources*, 5(5):47–52, October 1998.
9. A. Fox, S. D. Gribble, E. A. Brewer, and E. Amir. Adapting to Network and Client Variability via On-Demand Dynamic Distillation. In *Proceedings of the 7th International Conference on Architectural Support for Programming Languages and Operating Systems (ASPLOS)*, pages 160–170, Cambridge, Massachusetts, October 1996.
10. H. Garcia-Molina, T. Allan, B. Blaustein, R. M. Chilenskas, and D. R. Ries. Data-Patch: Integrating Inconsistent Copies of a Database after Partition. In *Proceedings of the 3rd Symposium on Reliability in Distributed Software and Database Systems*, pages 38–44. IEEE, October 1983.
11. J. Gray, P. Helland, P. O'Neil, and D. Shasha. The Dangers of Replication and a Solution. In *Proceedings of 1996 ACM SIGMOD Conference*, pages 173–182, June 1996.
12. H. Maruyama and K. Tamura and N. Uramoto. *XML and Java: Developing Web Applications*. Addison-Wesley, 1999.
13. J. S. Heidemann, T. W. Page, R. G. Guy, and G. J. Popek. Primarily Disconnected Operation: Experiences with Ficus. In *Proceedings of 2nd Workshop on the Management of Replicated Data*, pages 2–5, Monterey, CA, November 1992.
14. Y. Huang, P. Sistla, and O. Wolfson. Data Replication for Mobile Computers. In *Proceedings of 1994 ACM SIGMOD Conference*, pages 13–24, 1994.
15. T. Imielinski and B. R. Badrinath. Querying in Highly Mobile Distributed Environments. In *Proceedings of 18th Conference on Very Large Data Bases*, pages 41–52, August 1992.
16. J. Jones. private communication, 1998.
17. A. D. Joseph, J. A. Tauber, and M. F. Kaashoek. Mobile Computing with the Rover Toolkit. *IEEE Transactions on Computers: Special Issue on Mobile Computing*, 46(3), March 1997.
18. L. Kalwell Jr., S. Beckhardt, T. Halvorsen, R. Ozzie, and I. Grief. Replicated Document Management in a Group Communication System. In D. Marca and G. Bock, editors, *Groupware: Software for Computer-Supported Cooperative Work*, pages 226–235. IEEE Computer Society Press, 1992.
19. A. M. Keller, T. Ahamad, M. Clary, O. Densmore, S. Gadol, W. Huang, R. Razavi, and R. Pang. The DIANA Approach to Mobile Computing. In T. Imielinski and H. F. Korth, editors, *Mobile Computing*, pages 651–679. Kluwer Academic Press, 1995.
20. J. J. Kistler and M. Satyanarayanan. Disconnected Operation in the Coda File System. *ACM Transactions on Computer Systems*, 10(1):3–25, February 1992.
21. N. Krishnakumar and R. Jain. Protocols for Maintaining Inventory Databases and User Service Profiles in Mobile Sales Applications. In *Proceedings of the Mobidata Workshop*, October 1994.
22. H. Lei. *Uncovering and Exploiting the Intrinsic Correlations between File References*. PhD thesis, Columbia University, 1998.
23. H. Lei, M. Blount, and C. Tait. DataX: an Approach to Ubiquitous Database Access. In *Proceedings of 2nd Workshop on Mobile Computing Systems and Applications*, pages 70–79. IEEE, February 1999.
24. N. Cohen and Q. Zondervan. Specification of the COSMOS/MDSP Synchronization Protocol, Draft 0.31. IBM internal Design Document, January 1999.

25. B. D. Noble, M. Satyanarayanan, D. Narayanan, J. E. Tilton, J. Flinn, and K. R. Walker. Agile Application-Aware Adaptation for Mobility. In *Proceedings of 16th ACM Symposium on Operating Systems Principles*, pages 276–287, Saint Malo, France, October 1997.

26. Oracle Corporation. Oracle Lite Reviewers' Guide, 1998. http://www.oracle.com/mobile/olite/html/ol_review.pdf.

27. E. Pitoura and G. Samaras. *Data Management for Mobile Computing*. Kluwer Academic Publishers, 1998.

28. R. A. Merrick. DRUID - A Language for Marking-up Intent-based User Interfaces. IBM internal design document, March 1999.

29. Sybase, Inc. SQL Anywhere Studio - A Guide for Evaluation and Review, 1998. http://www.sybase.com/products/anywhere/sql_reviewers_guide.pdf.

30. C. Tait, H. Lei, S. Acharya, and H. Chang. Intelligent File Hoarding for Mobile Computers. In *Proceedings of 1st International Conference on Mobile Computing and Networking*, pages 119–125. ACM, November 1995.

31. The World Wide Web Consortium. Document Object Model (DOM) Level 1 Specification, W3C Recommendation 1-October-1998. http://www.w3c.org/TR/1998/REC-DOM-Level-1-19981001/.

32. The World Wide Web Consortium. XML Linking Language (XLL), W3C Working Draft 3-March-1998. http://www.w3c.org/TR/WD-xlink.

33. O. Wolfson, P. Sistla, S. Dao, K. Narayanan, and R. Raj. View Maintenance in Mobile Computing. *ACM SIGMOD Record*, 24(4):22–27, 1995.

Design and Evaluation of an Information Announcement Mechanism for Mobile Computers

Shigeaki Tagashira, Fumitake Inada *, Keizo Saisho, and Akira Fukuda

Graduate School of Information Science
Nara Institute of Science and Technology
8916-5 Takayama, Ikoma, Nara, 630-0101, Japan

Abstract. A mobile information announcement system has been constructed. In the system, a mobile computer can announce information at any place and any time. Moreover, the bandwidth problem between a mobile computer and an access point is pointed out. If the bandwidth of the network is narrow, all requests from clients cannot be processed by a mobile computer. An announcement mechanism for a mobile computer, which can effectively announce information with a narrow bandwidth network, is proposed. In this paper, in order to optimize performance of the proposed mechanism, parameters of the mechanism are analyzed in two kinds of networks through experiments. By optimizing parameters, the performance of the mechanism improves about 40% on wireless LAN compared with the performance without consideration of jitters in communication.

1 Introduction

The improvement of packaging and low power technologies makes mobile computers popular. Generally, a mobile computer is used to keep, generate, and modify personal information. The latest personal information usually exists on the mobile computer. Thus, it is important not only to get information but also to announce information from mobile computers. If a mobile computer can announce information, the latest personal information can be provided at any place and any time, and Live Broadcasting System from the outside and Handy-Phones through a network can be constructed by using mobile computers very easily. We have constructed and implemented an information announcement system for mobile computers, and confirmed the effectiveness of the system [1]. After this, we call entity of information "resource".

We discuss the bandwidth problem between a mobile computer and an access point on our system. A mobile computer can be connected to a network from anywhere using various interfaces. The difference in the interfaces influences the bandwidth and characteristics in communication. For example, ethernet has enough bandwidth to announce information, but wireless LAN has an insufficient bandwidth, and it is often interrupted by noise. If the bandwidth of a network

---

* Currently with Kyushu University, Fukuoka, Japan

H.V. Leong et al. (Eds.), MDA'99, LNCS 1748, pp. 135–145, 1999.
© Springer-Verlag Berlin Heidelberg 1999

is not sufficient, the traffic of announcing resources is limited, and it is also not effective for a mobile computer to announce concurrently all requested resources from clients while the traffic exceeds the limit of the network. All requested resources cannot be announced satisfying clients' needs.

Therefore, we have proposed an announcement mechanism which adopts the following policies designed to satisfy clients' needs within a given bandwidth as much as possible [2].

- A time-constrained resource is announced with high priority.
- A frequently accessed resource is provided as quickly as possible.
- The throughput of announcing resources is maximized.

In order to apply this mechanism to many kinds of networks and operating systems, the mechanism is implemented in the application layer. The policies are implemented by using the following operations in the mechanism.

- Traffic is controlled by limiting the number of concurrent operations for announcing resources.
- A resource is announced in order of its priority.

We have implemented a prototype of the information announcement mechanism and confirmed the effectiveness of it through experiments [2]. However, parameters which affect the performance in the proposed mechanism have not been evaluated in detail. The parameters need to be set to enhance the performance. In this paper, in order to optimize the parameters, the influences of the parameters are analyzed in two kinds of networks. By optimizing parameters, the performance of the mechanism improves about 40% on wireless LAN.

2   Overview of Our System

2.1   Structure of Our System

Fig. 1 shows an overview of our information announcement system for mobile computers. The system is based on the WWW (World Wide Web). In the system, we classify resources into the two types: one is the storage type that could be reused (ex. text data, image data, archive data, and so on) , and the other is the non-storage type that is only used one time (ex. time-constrained video and audio data). A mobile computer announces resources with suitable announcement ways according to their types. A mobile computer is associated with the Home WWW server that manages location of the mobile computer and keeps copies of storage type resources on the mobile computer. The copies are made on the Home WWW server when the mobile computer is connected. They are used to announce while the mobile computer is disconnected. They also work as cache during connection period. Therefore, traffic for storage type resources between a mobile computer and the Home WWW server only occurs when the coherence between original resources on the mobile computer and their copies on the Home WWW server is broken.

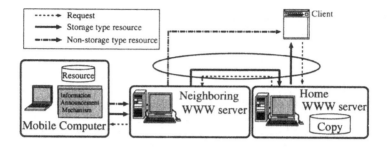

Fig. 1. Overview of Information Announcement System.

When a non-storage type resource is announced with the storage type announcement method, the following overheads are generated: (1)resources always go and come through the Home WWW server, and (2)copies, which are never used, are made on the Home WWW server. These overheads increase response time and reduce the throughput of non-storage type resources. It is important to shorten latency in a transmission route for a non-storage type resource. In order to alleviate these overheads, we introduced a Neighboring WWW server. The Neighboring WWW server is the nearest WWW server to the place where the mobile computer is connected. A non-storage type resource is provided for a client through a route of the mobile computer, the Neighboring WWW server, and the client. We consider the policy of announcing each type resource in order for the system to answer the following clients' needs.

Non-storage Type Resource: Non-storage type resources often include real-time resources. In addition, some resources are provided in the low quality if a time restriction cannot be satisfied. A client wants to get a non-storage type resource in the high quality. Therefore, non-storage type resources are announced prior to storage type resources.

Storage Type Resource: A copy of a storage type resource exists on the Home WWW server. For many resources, the coherence between the resource and its copy need not be completely kept. A client wants to get the latest resource. Therefore, a frequently accessed storage type resource should be provided as quickly as possible.

2.2   Information Announcement Mechanism

We show an information announcement mechanism in order to satisfy the clients' needs as much as possible even if a mobile computer is connected to a narrow bandwidth network. The mechanism consists of two mechanisms. One is the non-storage type resource announcement mechanism which announces non-storage type resources, and the other is the copy update mechanism which updates copies on the Home WWW server. When the bandwidth of the connected network is limited, the information announcement mechanism announces resources and updates copies according to the following policies:

- Non-storage type resources are announced prior to storage type resources.
- A frequently accessed storage type resource is updated prior to the other storage type resources.
- The throughput of updating is maximized.

In order to realize the above mechanism, we focus on the copy update mechanism since the mechanism affects the throughput of announcing non-storage type resources and total time of updating the copy. The non-storage type resource announcement mechanism and the copy update mechanism can cooperate on announcing resources. Announcing non-storage type resources and updating copies of storage type resources should be achieved by the following operations.

- Traffic is controlled by limiting the number of connections for announcing or updating resources.
- A resource is announced in order of its priority.

The optimal number of connections is decided according to the following conditions:

(1) Condition for announcing a non-storage type resource:
     A sufficient bandwidth given to the requested resource is ensured.
(2) Condition for updating a copy of a storage type resource:
     The total throughput of updating copies is maximized.

If condition (1) is not satisfied, condition (2) is ignored. When no non-storage type resource is announced, only condition (2) is available. We consider implementation of the proposed mechanism at three levels: (1)Hardware Level[3], (2)Kernel Level[4], and (3)Application Level. Since a mobile computer moves to various places, it will be connected to various networks. It is important for mobile computers to use the mechanism in many networks. Therefore, we adopt the application level.

3   Implementation of Copy Update Mechanism

3.1   Structure of Copy Update Mechanism

The structure of the copy update mechanism is shown in Fig. 2. We define three terms for the mechanism.

Number of Connections (NC): NC is the current number of connections used to update copies.
Estimated Optimal Number of Connections (EOPNC): EOPNC is the number of connections expected to maximize throughput of updating copies. **EOPNC** is estimated from current throughput.

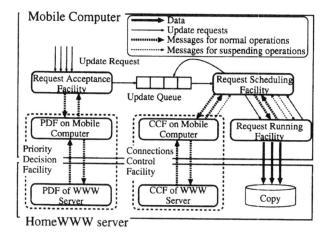

Fig. 2. Structure of Copy Update Mechanism.

Optimal Number of Connections (OPNC): OPNC is the number of connections which actually maximizes throughput of updating copies.

This mechanism consists of one update queue and five modules.

Update Queue (UQ): This queue holds update requests. An update request is issued from a user or caused by suspending update.

Priority Decision Facility (PDF): Priorities are assigned to storage type resource according to the importance. This facility decides priority of a storage type resource. The copies of resource are updated in the order of descending their priorities. The chance for a client to get the latest resource depends on priority of the resource. We introduce A/U. A/U is the ratio of access frequency to update frequency.

Request Acceptance Facility (RAF): This facility stores an update request in **UQ**. The position of the request in **UQ** is decided by its priority. The priority of the request is given by **PDF**.

Connections Control Facility (CCF): This facility decides **EOPNC** according to communication condition described in Sec. 3.2.

Request Scheduling Facility (RSF): If **UQ** is not empty and it is possible to make a new connection, **RSF** asks **RRF** to process the request at the top of **UQ**.

Request Running Facility (RRF): This facility processes an update request from **RSF**. **RRF** sends the requested resource to the WWW server.

The bold solid arrows, the thin solid arrows, the bold dotted arrows, and the thin dotted arrows show flow of resource, the flow of update requests, the flow of messages for normal operations, and the flow of messages for suspending operations, respectively.

While the system updates copies, **EOPNC** may change. Therefore, **CCF** sends **RSF** the message which notifies to change **NC** when **EOPNC** changes. **RSF** processes the messages according to it's type.

Increase of connections: **RSF** only increases **EOPNC**. Because **NC** increases when it is lower than **EOPNC** and there are any requests in **UQ**.

Decrease of connections: When **NC** is larger than or equal to **EOPNC**, **RSF** selects the connection which has the lowest priority among current connections, and asks **RRF** to suspend the connection in order to decrease **NC**. **RRF** suspends the asked connection, and then, **RSF** returns the suspended request to **UQ**.

3.2   Connections Control Facility (CCF)

CCF administrates connections for update and decides **EOPNC**. This facility is partitioned into two modules, and they are set to a mobile computer and its Home WWW server. The part sets to the Home WWW server observes the throughput, and send it to its partner on a mobile computer. The partner decides whether **NC** to change or not using received information. When **EOPNC** changes, it also sends the changing message to **RSF**.

Before describing the decision algorithm, we explain the parameter of this algorithm.

α: This parameter is a threshold for changing **EOPNC**. The value is also a margin for spatial suppression of the influence of the jitters in communication. When the value is too small, changing messages are frequently issued. Although the influence of the jitters can be reduced using a large value of α, it is impossible to reach **OPNC**.

M: This parameter is a margin for temporal suppression of the influence of the jitters in communication. When the value is too small, **EOPNC** is affected by the jitters in communication. The influence of the jitters becomes large and **EOPNC** is frequently changed. However, when the value is too large, it takes long time to reach **OPNC**.

We show the conditions to change **EOPNC**.

(1) Condition of Non-storage type resource:
$$TH_R < TH_C$$
TH_R is the requested throughput. TH_C is the measured throughput.

(2) Condition of Storage type resource:
$$-\alpha < R - 1 < +\alpha$$
R is ratio of the current throughput to the last time throughput.

Fig. 3 shows the **EOPNC** decision algorithm. There are three states:

State(a): Condition (1) is not satisfied,
State(b): Condition (2) is not satisfied caused by increase of throughput, and
State(c): Condition (2) is not satisfied caused by decrease of throughput.

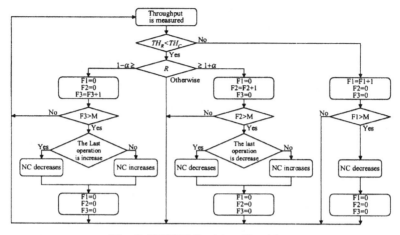

Fig. 3. EOPNC Decision Algorithm.

Counter F1, F2, and F3 correspond to State (a), (b), and (c), respectively. When the same state repeats M times, **EOPNC** is changed according to states and the last changing operation.

4   Evaluation

We implement the proposed information announcement mechanism on our system described in Sec. 2 and evaluate it. The aim of this evaluation is to analyze the influence of α and M as described in Sec. 3.2. and to decide the optimal ones.

In order to obtain the target values of the throughput of the non-storage type resource and the total update time of the storage type resources in ethernet and wireless LAN, multiple storage type resources and one non-storage type resource are sent simultaneously with fixed **NC**. The total update time of storage type resources and throughput of the non-storage type resource are measured. Table 1 shows other parameters on the experiment. Fig. 4 shows the result of the experiment. The target update time in ethernet and wireless LAN are about 65 seconds and 97 seconds, respectively.

4.1   Total Update Time

The total update time of storage type resources and the throughput of the non-storage type resource using the proposed mechanism are measured by varying α and M on the same environment as the above experiment. Fig. 5 shows the total update time.

In addition, we investigate the influence of α and M from the changes of **EOPNC**. Fig. 6 shows the result. In ethernet, the optimal value of α is 0.2.

Table 1. Parameters of experiments.

Network media	Ethernet	Wireless LAN
Speed (bps)	10M	1M
Size of storage type resources (bytes)	1M	50K
Number of storage type resources	40	
Requested throughput of a non-storage type resource (bytes/s)	8,000	
Initial number of connections	1	

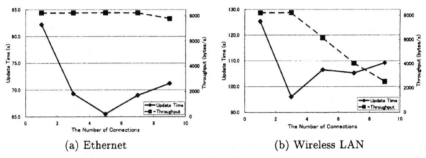

(a) Ethernet   (b) Wireless LAN

Fig. 4. Fixed NC vs. Update Time and Throughput.

When the value of α is too small, **EOPNC** is frequently changed by the jitters in communication. However, when the value of α is too large, **EOPNC** is not changed despite a chance for changing **OPNC**. The changes of **EOPNC** are reduced. The total update time is affected by the value of α more than that of M. The influence of M is very little at the optimal value of α. The total update time at the optimal value of α and M is about 67 seconds. This value is near to the target one (65 seconds).

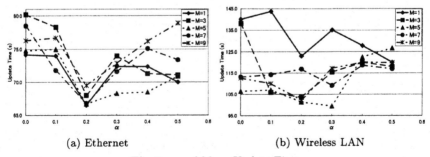

(a) Ethernet   (b) Wireless LAN

Fig. 5. α and M vs. Update Time.

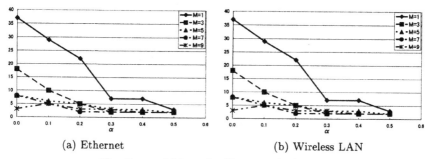

(a) Ethernet   (b) Wireless LAN

Fig. 6. α and M vs. the Changes of EOPNC.

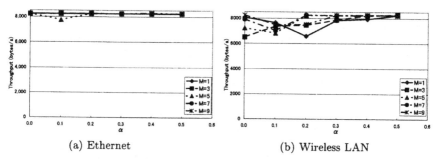

(a) Ethernet   (b) Wireless LAN

Fig. 7. α and M vs. Throughput.

In wireless LAN, the total update time is affected by both α and M. The optimal value of α and M is 0.3 and 5, respectively. The behavior of α is the same as that of ethernet. The behavior of M is, however, different from that of ethernet because the influence of the jitters in wireless LAN is very much and the jitters must be suppressed by both α and M. The total update time at the optimal value of α and M is about 100 seconds. This value is near to the target one (97 seconds). The value is 40% less than the value with no margin (α is 0.0 and M is 1).

The total update time at the optimal value of α and M is near to the target one in both networks.

As the result, in ethernet, the optimal value of α is 0.2 and that of M is 5. In wireless LAN, the optimal value of α is between 0.2 and 0.3 and that of M is 5. The values measured by using the proposed mechanism can get on toward the target values with the optimal value of α and M.

4.2   Throughput of Non-storage Type Resource

We evaluate the throughput of the non-storage type resource in the experiment. Fig. 7 shows the result. In ethernet, the proposed mechanism can always satisfy

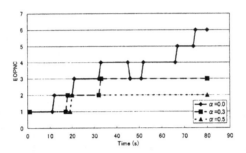

Fig. 8. Transition EOPNC in Wireless LAN.

the request throughput. In this case, **EOPNC** is around four. If each connection is assigned to the bandwidth equally, the throughput of a connection is more than 75Kbytes per second. This value is too large compared with the request throughput of the non-storage type resource (8Kbytes per second). In wireless LAN, when the value of α is large, the request throughput can be satisfied. However, when the value of α is small, the request throughput cannot be satisfied because of the influence of changing a connection. In order to investigate the influence of α, we measure the transition **EOPNC**. In the experiment, network media is wireless LAN and the value of M is 5. Fig. 8 shows the result. When the value of α is optimal (0.3), **EOPNC** converges at **OPNC**. Although **EOPNC** cannot reach **OPNC** when the value of α is 0.5, **EOPNC** is set to stable value and the value is smaller than **OPNC**. Therefore, the non-storage type resource is assigned to the requested throughput. When the value of α is 0.0, **EOPNC** is unstable changing and out of **OPNC**. Therefore, the non-storage type resource is not assigned request throughput.

5   Conclusion

In this paper, the bandwidth problem between a mobile computer and an access point has been discussed, and the information announcement mechanism has been proposed. The mechanism effectively announces the resources even if a mobile computer is connected to a narrow bandwidth network. The mechanism limits traffic of storage type resources in order to announce non-storage type resources prior to storage type resources. Traffic is controlled by limiting the number of connections for announcing storage type resources. We have implemented the prototype and evaluated it. In order to optimize α and M which are parameters of the mechanism, we have analyze the influence of them in two kinds of networks. By optimizing parameters, the performance of the mechanism improves. The followings are future works:

- We construct practical killer applications by using the proposed mechanism.
- Priority Decision Facility (**PDF**) can be flexible in the policy so that the system can be adapted to many kinds of clients' needs.

References

1. S. Tagashira, K. Nagatomo, K. Saisho, and A. Fukuda: "Design and Evaluation of a Mobile Information Announcement System Using WWW," Proc. the IEEE Third Int. Workshop on Systems Management (SMW'98), pp.38–47, 1998.
2. S. Tagashira, K. Saisho, F. Inada, and A. Fukuda: "A Copy Update Mechanism for a Mobile Information Announcement System — Transmitting Non-storage Type Resources with Redundancy —," Proc. Int. Workshop on Data Warehousing & Data Mining, Mobile Data Access, and New Database Technologies for Collaborative Work Support & Spatio-Temporal Data Management (DWDM/MDA/NewDB) , pp.261–272 (1998. 11).
3. M.de Prycker: "Asynchronous Transfer Mode-solution for broadband ISDN-," Ellis Horwood, 1991.
4. Sally Floyed and Van Jacobson: "Link-sharing and Resource Management Models for Packet Networks," IEEE/ACM Trans. on Networking, Vol.3 No.4, pp.365–386, 1995.

Active Rule System
for Adaptive Mobile Data Access [*]

Shiow-yang Wu   Shih-Hsun Ho

Department of Computer Science and Information Engineering
National Dong Hwa University
Hualien, Taiwan, R.O.C.
showyang@csie.ndhu.edu.tw,

Abstract. We advocate the employment of active rule systems for adaptive mobile information services. We propose a modular framework and a mobile rule processing technique with database connectivity. The technique combines static and dynamic analysis to uncover data access semantics of a rule program which facilitates intelligent caching and prefetching to conserve bandwidth, reduce processing cost, and support disconnected operations. We devise a performance model to compare our approach with traditional approach. Trace-driven simulation successfully demonstrates the feasibility and performance improvement.

1   Introduction

Mobile computing is characterized by limited resources, low bandwidth, and dynamically changing environment[1] which offers a great opportunity for applying intelligent techniques toward the provision of responsive information services. As a motivating example, imagine the scenario in which a mobile user on a moving vehicle is approaching a mountain area where radio signal is expected to be weak or completely blocked. According to the current traffic condition and also from the speed and direction of the vehicle, no mobile support station will be available for the next half an hour. For assisting the user who must finish her work for an important business meeting, the underlying information system has several choices. The easiest way is to take no action and run the risk of sudden disconnection, unfinished work, and a failed business trip. More preferably, a highly adaptive system can sense the current situation and take the initiative in preparing for disconnected operation. Our goal is to develop intelligent techniques for building responsive and effective mobile information systems. In particular, we advocate the employment of active rule system for intelligent adaptation.

Rule systems have been successfully used in providing many desirable features for traditional database systems [13]. The active rule component effectively turns a passive database into one that can react dynamically to status changes and respond actively with rule inference mechanism. For mobile data management, we

---

[*] This work was supported in part by National Science Council under project number NSC 88-2213-E-259-001.

H.V. Leong et al. (Eds.), MDA'99, LNCS 1748, pp. 146–155, 1999.

found the active rule concept to be a powerful tool for intelligent adaptation and proposed an adaptive mobile information system framework in an earlier paper [14]. In this paper, we present the underlying rule processing technique which enables effective rule execution under various resource constraint and operating conditions. We propose a mobile data management and rule execution strategy with intelligent caching and prefetching to facilitate seamless information processing regardless of the connection status. We devise a performance model to compare our approach with traditional on-demand approach. Trace-driven simulation results successfully demonstrate the feasibility of our approach and the potential for significant performance improvement.

The rest of the paper is organized as follows. Section 2 provides a survey of related work. Section 3 presents our framework for mobile rule processing. In Section 4, we introduce the intelligent caching and prefetching technique. A performance model is introduced in Section 5 to analyze the effectiveness of the proposed techniques. In Section 6, we discuss a trace-driven simulation method and the results of various experiments. Section 7 concludes the paper.

2   Related Work

The need for intelligent adaptation is considered essential for mobile data management [5]. Similar ideas have been discussed under terms like context-aware [11], application-aware [10], environment-directed [12], and adaptive [3] information systems. The decision on when and how to adapt can be made by either the underlying system (*application-transparent adaptation*) or the application programs (*application-aware adaptation*) [9]. Application-transparent implies that no change is needed to existing applications which also means no application-specific feature can be exploited. On the other hand, the performance and flexibility offered by the application-aware adaptation may very well come with higher complexity and software development cost. Our framework integrates existing rule system and database systems for intelligent adaptation which offers the benefits of both application-transparent and application-aware adaptation.

A database system with an integrated rule system is called an *active database system*[13]. A rule system is composed of a *working memory*, a set of *rules*, and an *inference engine*. Working memory is a global space composed of data objects called *working memory elements* (WMEs) representing the system state. A rule is a condition-action pair. The inference engine provides a three-phase cyclic execution model of *condition evaluation, conflict-resolution* and *action firing*. A rule with a set of WMEs satisfying the conditions is called an *instantiation*. The set of all instantiations constitutes the *conflict set*. Conflict-resolution selects one instantiation from the conflict set for firing. Firing an instantiation executes the actions which may add, delete, or modify WMEs in the working memory. The cycle repeats until no more rule can be fired. Rules in active database systems are normally enhanced with event triggering capability and known as *event-condition-action rules* (ECA rules)[2, 7]. An ECA rule is triggered on the occurrence of certain events. The condition part of the rule is evaluated against the current state to determine whether the conditions are satisfied. The action part of the rule is then executed whenever the conditions are matched.

ECA rules are natural candidate for adaptive mobile information access. Resource and environment changes can be modeled as events. The condition evaluation, action firing, and rule inference mechanisms can be used for making proper adaptation decision. Our main contribution is to provide a framework and the underlying rule processing techniques for building mobile information systems that can respond actively to resource and environment changes.

3   A Framework for Adaptive Mobile Information Access

We describe our framework in brief. Readers are referred to [14] for more information. As depicted in Figure 1, we proposed the integration of a *mobile event engine*, a *rule system*, and a *database system* for adaptive information access. The mobile event engine is to detect status changes and to trigger the rule system. In additional to the database and temporal events [13], we suggest the inclusion of mobility, resource, and environment events. Whenever the status of a mobility variable (location, speed, etc.), a resource or an environment variable has changed to a critical level, a corresponding event occurs. Such changes are detected by the *event detector* and treated as *primitive events* which can be combined to form *composite events*. The *event delivery module* transforms the events into *event elements* which are in the same format as WMEs of the rule system. In this way, the event engine can be easily integrated with the rule system.

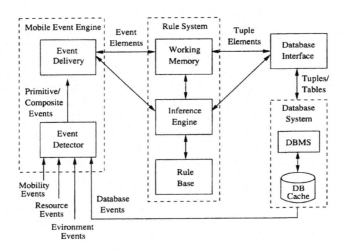

Fig. 1. A Modular Framework for Building Adaptive Mobile Information Systems

The *database interface* integrates rule and database systems. Tuples from the database are converted into *tuple elements* for the rule system. Data generated within the rule system can be saved into database. By conforming to standard, the interface can interact with multiple and heterogeneous databases.

In this paper, we focus on the mobile rule processing technique which is designed based on the following observations:

- Most rule programs execute in phases. During each phase, only a small subset of the rules are active. These rules exhibit strong data access locality that can be determined by examining the condition part of the rules.
- It is common to use a *goal element* to control phase execution. All rules of the same phase match the same goal element. *Phase changing rules* are used to update the goal elements for switching between phases.
- The execution of a rule results in small changes to the database. An active rule which is not selected for firing in the current cycle is likely to remain valid in the next cycle since most data remain unchanged. Therefore, it is beneficial to keep the data available through caching.

Our framework is organized as an integration of static and dynamic analysis for prefetching and caching as depicted in Figure 2. The phase execution behavior and strong locality of data access suggest a strategy that prefetch and cache data according to the phase execution semantics. Static analysis can uncover the phase structure and partition the rules into clusters. The purpose is to group together rules with similar data access pattern. The data accessed by the rules in a cluster can be determined by analyzing the condition part of these rules. The result of the static analysis is the rule clusters and a set of *view definitions* corresponding to the data accessed by each cluster. These information are used by the cache manager to determine when and what data to prefetch and retain in the cache. Dynamic analysis is to monitor rule execution and to adjust prefetching and caching policy if necessary. User profiles are used whenever applicable to improve the effectiveness of the prefetching and caching decisions. Details of our technique will be presented in the next section.

4   Predictive Caching and Prefetching

Phase structure analysis described above can be done by grouping together all rules that match the same goal element or by using *data dependency graph* [4] and *clustering* techniques [6]. The set of view definitions extracted for a rule cluster is formulated to include all data that can possibly be accessed by any rule in the cluster. User profile is applied to limit the scope and range.

We apply the technique on the following rules. Continuing with the example in Section 1, the mobile user is about to report her reorganization plan to the CEO at the headquarter. For an unexpected last minute change, part of the proposal must be modified along the way. Reorganization planning strategy is determined based on bandwidth. In normal condition, a detail planning can be taken. In case that the bandwidth falls and the time remaining is short, a simple salary cut strategy is adopted. We also assume that the manager has an authority level to examine employee records with salary lower than 100,000 dollars.

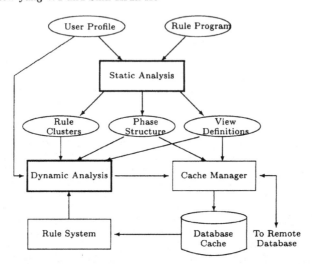

Fig. 2. Integration of Static and Dynamic Analysis for Mobile Rule Processing

```
rule Strategy_Selection {
// A phase changing rule to select a simple strategy if the
// available bandwidth is low and the time remaining is short.
on (bandwidth < 1 Mbps) AND (time_remain < 30 minutes)
if (g:Goal)  // a goal element exist
then
    g.current_goal = ADJUST_SALARY }

rule Salary_Cut {
// Cut 10% off the salary of high-paid employee (salary > 80000)
// for department with negative earnings.
if (g:Goal:: g.current_goal == ADJUST_SALARY)
    (d:Department:: d.earnings < 0)
    (e:Employee:: e.department == d.name & e.salary > 80000)
then
    e.salary = e.salary - e.salary/10 }

rule Salary_Raise {
// Raise 10% the salary of low-paid employee (salary < 10000)
// for department with earnings > one million.
if (g:Goal:: g.current_goal == ADJUST_SALARY)
    (d:Department:: d.earnings > 1000000)
    (e:Employee:: e.department == d.name & e.salary < 10000)
then
    e.salary = e.salary + e.salary/10 }
```

During the phase structure analysis, all ADJUST_SALARY rules will be grouped together into the same partition. The data requirement can be extracted from the test conditions and represented by the following view definitions.

```
SELECT *
FROM    Department
WHERE   earnings < 0 OR earnings > 1000000

SELECT *
FROM    Employee
WHERE   salary > 80000 OR salary < 10000
```

A closer look at the rules reveals the disjointness of the test conditions which separates the rules into two clusters. The view definitions becomes

```
SELECT *
FROM    Department
WHERE   earnings < 0

SELECT *
FROM    Employee
WHERE   salary > 80000
```

for the first cluster consisting of the Salary_Cut rule, and similarly for the second cluster consisting of the Salary_Raise rule. The range of data can be further constrained by considering the manager's authority level:

```
SELECT *
FROM    Employee
WHERE   salary > 80000 AND salary < 100000
```

Only the data in the views are needed for the rule cluster to execute properly which significantly reduces the volume of data transfer and conserves valuable bandwidth. We now describe our mobile rule execution strategy.

- Apply static analysis to partition the rules and extract view definitions.
- Upon execution of a phase, prefetch all views of that phase in one connection.
- Log all updates. At the end of a phase, send the net updates back to the remote database in one connection.
- Upon disconnection, simply use the data in the cache to proceed normally. As long as all views have been cached, the execution remains valid.

Intuitively, our approach improves latency by keeping currently used and to be used data nearby. Bandwidth is conserved by loading only the necessary views. Disconnected operations are supported by prefetching and caching data for the current and subsequent phases. In later sections we will demonstrate that this simple approach can result in significant saving in rule execution cost.

5 Performance Model and Comparison

We devise a performance model and compare our approach with traditional approach which accesses data on-demand. We assume that each request results in a single unit of data access. The model is described by the following parameters:

C_c Cost for setting up a connection.
C_t Unit cost of data transmission.
C_{io} Unit I/O cost.
R_p Cache hit ratio (with prefetching).
R_{np} Cache hit ratio (w/o prefetching).
T Total number of data requests.
V_p Total units of prefetched data.
H The additional hits results from the prefetching of V_p (i.e. $T*(R_p-R_{np})$).

A cache hit can be satisfied locally. A cache miss incurs the cost of connection setup, remote access, and data transfer. The total cost for the approach with and without prefetching (E_p and E_{np}) can be modeled as:

$$E_{np} = T * R_{np} * C_{io} + T * (1 - R_{np}) * (C_c + 2C_{io} + C_t)$$
$$E_p = (C_c + (2C_{io} + C_t) * V_p) +$$
$$\quad\quad T * R_p * C_{io} + T * (1 - R_p) * (C_c + 2C_{io} + C_t)$$

where the first item of E_p is the cost of prefetching. For the new approach to be effective, we must have $E_{np} - E_p > 0$. By simple algebraic manipulation, we obtain the relationship between E_{np} and E_p as follows.

$$E_{np} - E_p = (H - 1)C_c + (H - 2V_p)C_{io} + (H - V_p)C_t$$

This can be depicted in Figure 3 which shows the desired level of accuracy in prefetching. Region A ($H > 2V_p$) represents the case when prefetching will definitely have an advantage. Region B ($2V_p > H > V_p$) stands for the situation that prefetching does not guarantee but highly likely to provide better performance. This is because $C_t \gg C_{io}$ in mobile environment. If C_{io} can be safely ignored, then prefetching is beneficial in both region A and B. It is not cost worthy in region C, which means when the additional hits is smaller than the prefetched units ($H < V_p$), the overhead prevails. Regions D, E, and F are cases when none of the data prefetched is actually used which are very unlikely.

6 Performance Evaluation

We have devised a trace-driven simulation method as depicted in Figure 4. The CLIPS rule system [8] was used to generate the execution trace. The test programs are scalable rule programs that access databases of arbitrary size generated by a data generator. The trace files are produced by running the rule programs on CLIPS over databases of increasing size. Each trace file contains the detail record of every single rule firing and the data accessed by that rule.

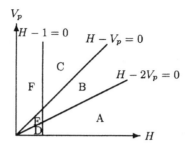

Fig. 3. Effectiveness of prefetching vs. no prefetching

Static analysis is applied to produce the rule clusters and view definitions. The simulation program takes a trace file, the rule clusters and view definitions as input and produces detail figures of the execution cost under various settings. Given a set of parameters such as the cache size, C_c, C_t, C_{io}, and the replacement policy, the simulation program parses the trace file and performs a pseudo execution of the rule program to accumulate all relevant costs. We are interested in connection setup, I/O, and transmission cost. The connection cost is important since each connection consumes valuable power and bandwidth. I/O and transmission cost reduction improves response time and saves energy as well.

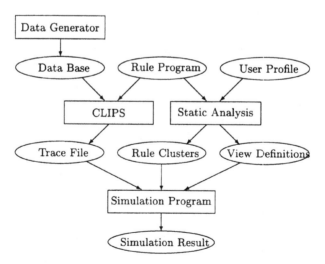

Fig. 4. Simulation Method

Figure 5, 6, and 7 present the comparison between on-demand approach and our scheme with small(160), median(2560), and large(40960) disk cache units. Our approach is superior in all three dimensions. Take Figure 5 as an example, our approach incurs much lower connection cost. This is because we need only one connection to prefetch all data required in a particular phase. When the cache is large enough(to hold the largest view), the number of connections required is exactly the number of phase changes during the program execution.

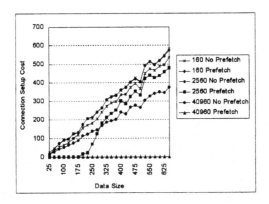

Fig. 5. Connection cost comparison

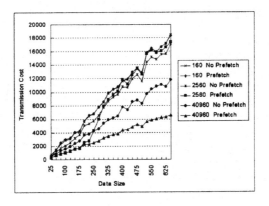

Fig. 6. Transmission cost comparison

7   Conclusions and Future Work

We presented a framework and techniques for building adaptive mobile information systems. The techniques employs semantic analysis to predict future data requirement. Significant cost saving can be achieved if the data is prefetched and cached at the right time. Locally cached data also facilitate disconnected operations. To complete the design, a mobile event engine is necessary. We are working on the event subsystem as well as its integration with other modules.

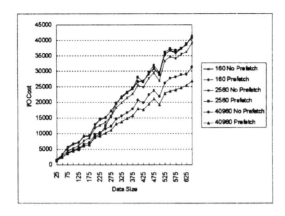

Fig. 7. I/O cost comparison

References

1. G. Forman and J. Zahorjan. The challenges of mobile computing. *IEEE Computer*, 27(6):38–47, Apr. 1994.
2. E. N. Hanson and J. Widom. An overview of production rules in database systems. *The Knowledge Engineering Review*, 8(2):121–143, June 1993.
3. T. Imielinski and S. Vishwanathan. Adaptive wireless information systems. Technical report, Department of Computer Science, Rutgers University, 1994.
4. T. Ishida and S. J. Stolfo. Toward the parallel execution of rules in production system programs. In *IEEE Intl. Conf. on Parallel Processing*, pages 568–574, 1985.
5. R. H. Katz. Adaptation and mobility in wireless information systems. *IEEE Personal Communications*, 1(1):6–17, 1994.
6. D. P. Miranker, C. Kuo, and J. Browne. Parallelizing compilation of rule-based programs. In *IEEE Intl. Conf. on Parallel Processing, Vol. II*, pages 247–251, 1990.
7. N. W. Paton, editor. *Active Rules in Database Systems*. Springer, New York, 1999.
8. G. Riley. CLIPS: A tool for building expert systems. (http://www.jsc.nasa.gov / clips/CLIPS.html).
9. M. Satyanarayanan. Mobile information access. *IEEE Personal Communications*, 3(1), Feb. 1996.
10. M. Satyanarayanan, B. Noble, P. Kumar, and M. Price. Application-aware adaptation for mobile computing. *Operating System Review*, 29, Jan. 1995.
11. B. Schilit, N. Adams, and R. Want. Context-aware mobile applications. In *IEEE Workshop on Mobile Computing Systems and Applications*, Santa Cruz, CA, U.S., Dec. 1994.
12. G. Welling and B. R. Badrinath. Mobjects: Programming support for environment directed application policies in mobile computing. In *ECOOP'95 Workshop on Mobility and Replication*, Aug. 1995.
13. J. Widom and S. Ceri. *Active Database Systems: Triggers and Rules for Advanced Database Processing*. Morgan Kaufmann, San Francisco, California, 1996.
14. S. Wu and C.-S. Chang. An active database framework for adaptive mobile data access. In *Workshop on Mobile Data Access, 17th International Conference on Conceptual Modeling, Singapore*, 1998.

Preserving Smooth Traffic and High Presentation QoS for VBR-Encoded Video

Se-Jin Hwang[1] and Myong-Soon Park[2]

[1] LG Information and Communications Ltd., Korea
hsj@lgic.co.kr
[2] Computer Science Dept., Korea University, Korea
myongsp@cslab5.korea.ac.kr

Abstract. This paper proposes a feedback based rate control mechanism to achieve both smooth traffic and the high presentation quality of encoded video. We observed that the aggregated required bandwidth of fluctuating consumption rate may be zero in time scale longer than negotiated at session establishment. We use such a notion in controlling rate. In our mechanism, optimism is found in that we do not make immediate response to high consumption rate. Traffic would not need to be changed by feedback if total aggregated bandwidth until next service round is smaller than the amount of resources negotiated at channel establishment. We show that rate control based on such an optimism guarantees acceptable QoS compared to previous mechanisms.

1 Introduction

Rate control for multimedia communication with encoded video should consider two conflicting goals: high presentation quality, smooth traffic[1–5]. Since these goals has tradeoff relationship, high intelligence needs to be assisted.

Recent trends to achieve smooth traffic of encoded video is to characterize the traffic during more than two intervals. It is based on the fact that aggregating actual bandwidths during several intervals might be smaller than average bandwidth negotiated at channel establishment. RED−VBR(**RE**negotiation Deterministic-**VBR**)[3] implements such a notion by building set of (bandwidth, interval) pairs before networking. These informations are computed in advance of service by scanning streams. Adjusting dynamically the pairs in service time depending on the burstiness was developed[4, 5]. To service video stored in disks, Salehi proposed an rate control mechanism considering the resource in client node[2].

The shortcoming of these techniques is to bound actual bandwidth of video to be within supportable bandwidth candidates. Therefore, it is unavoidable to send feedback to control production rate or to degrade the presentation quality if the actual bandwidth would be greater than one supportable. Salehi's approach[2] assumes that sender has information of client resource, hence, it is not proper to usual network environment.

H.V. Leong et al. (Eds.), MDA'99, LNCS 1748, pp. 156–165, 1999.

In this paper, we propose a rate control mechanism based on feedback, which optimistically deals with fluctuating consumption rate. The fundamental idea is similar to RED−VBR series approaches[3–5]. However, we expect no packet loss even if actual bandwidth exceeds the bound of current service round, since aggregated bandwidth at next rounds coule be within the bound. Such a notion leads us to make a conclusion that we do not need to promptly send feedback message even after the detection of rate fluctuation. Optimistic rate control mechanism insensitively reacts to fluctuating rate compared to previous mechanisms. By doing it, the production rate of source is not changed and traffic is kept being rather static than the previous mechanisms. It offers us with the smooth traffic, and does not hurt visual quality. We implemented our method into practical transport layer for multimedia networking, **MuSCLE** (**Mu**ltimedia **S**tream **C**ommunication **L**ayer **E**nhancement) by modifying Linux kernel. It is optimized for multimedia networking purpose.

The balance of this paper is organized as follows. Section 2 describes the flow model of feedback-based rate control, and simply introduces the basic idea of our mechanism[1] Section 3 describes our mechanism, optimistic rate control, and then its performance is analyzed in section 4. We make a conclusion in section 5.

2 Flow Specification and Basic Idea

Flow is characterized by the number of packet in sink buffer(a buffer at sink node) at the t-th service round , and it is denoted as follows:

$$S_t = \sum_{k=1}^{t} \rho_k \times \gamma + P_p, (|\rho_x| < R_b, x = 1, 2, \ldots, t) \tag{1}$$

Both source and sink are activated each service round, γ. The number of packet in sink buffer at the t-th round is computed by $\rho_t \times \gamma$. ρ_t is the difference between production rate of packet(P_r^t) and consumption rate at sink(C_r^t). If $P_r^t > C_r^t$ at the t-th round, ρ_t is positive. Otherwise, ρ_t is negative. P_r^0 is same to R_b. Sink is aware of R_b at channel establishment as average bandwidth. P_p indicates the number of packet prefilled in sink buffer to avoid buffer underflow from the start.

Fluctuation is detected by the coverage of S_t in sink buffer. Sink buffer has a pair of threshold denoted as { L_b , H_b }. The size of the sink buffer is denoted as B_s. L_b is a lower bound and H_b a higher bound respectively. If $S_t \in [L_b, H_b]$, flow is *stable*. However, If $S_t \in [0, L_b)$, buffer may be underflowed at the $(t+1)$-th round. $S_t \in (H_b, B_s]$ may cause buffer overflow. Feedback message is sent in order to prevent these two situations. P_r^{t+1} , C_r^t and λ_t can be defined as follows:

$$C_r^t = R_b + \rho_t, (|\rho_t| < R_b) \tag{2}$$

---

[1] In this paper, we use term 'RED−VBR' as representative mechanisms with strict bound checking.

$$P_r^{t+1} = min(max(R_b + \lambda_t, 0), \frac{A_{t+1}}{\gamma}, L_{max}) \tag{3}$$

$$\lambda_t = f(\rho_t, S_t, L_b, H_b, B_s) \tag{4}$$

A_t is the number of packet in source buffer at the t-th round and L_{max} is the link capacity. Function f in Eq. (4) is employed in sink node. It decides whether λ_t should be sent or not. Arguments in f is QoS parameters. Source has a flow regulator composed of min, max operators in Eq. (3). S_t can be broken down into a series of pairs ($S_{(start,end)}$, λ_{end}) as follows:

$$S_t = (S_{(1,k+1)}, \lambda_{k+1}) + (S_{(k+2,v)}, \lambda_v) + \dots, (S_{(s,e)} = \sum_{k=s}^{e} \rho_k \times \gamma) \tag{5}$$

For more simplicity, there is no overlapping between P_r^{t+1} and λ_t at any t-th round. Under such a flow model, Eq. (6) formalizes our idea.

$$\left((S_{(t,t+k)} < 0 \wedge S_{(t+k+1,t+p)} > 0) \vee (S_{(t,t+k)} > 0 \wedge S_{(t+k+1,t+p)} < 0) \right) \approx 0 \tag{6}$$

The signs of both $S_{(t,t+k)}$ and $S_{(t+k+1,t+p)}$ contribute to Eq. (6) since these have different signs each other. If there is no packet loss in $S_{t+k} \in [0, L_b)$, we do not send λ_{t+k}. We predict $S_{(t+k+1,t+p)} > 0$ from next service round. We refer to it as the *self-stability* of flow. If $S_{(t+k+1,t+p)} < 0$, the discontinuity of screen will appear, that is, presentation quality will suffer. Meanwhile, if $S_{(t+k+1,t+p)} > 0$, traffic is kept being smooth without λ_{t+k}, further the degradation of presentation quality does not exist.

3 Optimistic Rate Control

This section addresses optimistic rate control mechanism, and compares it with RED−VBR mechanism, most recently proposed strategy to meet QoS requirement of encoded video.

3.1 Qos Parameter

Optimistic rate control predicts ρ_{t+1} with a different sign from ρ_t's. Such a prediction might incur either packet loss or screen glitches if the signs of both ρ_t and ρ_{t+1} are identical. It is necessary to bound the loss rate given by user. QoS parameters required for optimism are $\rho_{max(t+1)}^e$ (future maximum required bandwidth), τ (maximum loss rate). In this paper, we define these two QoS parameters respectively as follows:

$$\rho_{max(t+1)}^e = \frac{R_b}{2RHO}, \tau = \frac{\rho_{max(t+1)}^e}{2TAU} \tag{7}$$

By specifying RHO and TAU, we can control the degree of QoS. Also, we limit the maximum number of prediction of *self-stability* with MXC, which is refered to as maximum prediction bound.

3.2   Buffer Organization and Algorithm

To guarantee that packet loss is preserved under $\tau \times \gamma$, we use one more thresholds in the buffer, usually it is known to be a *bounds*. The threshold is placed as far as $(\rho^e_{max(t+1)} - \tau) \times \gamma$ from H_b and L_b. Buffer orgnization is illustrated in Figure 1 compared to RED$-$VBR.

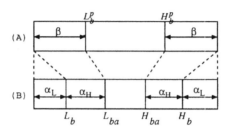

Fig. 1. Buffer organizations in both mechanisms((A):RED$-$VBR,(B):optimistic rate control)

Two thresholds in optimistic rate control are denoted as: $D^l = \{L_b, H_b\}$, $D^h = \{L_{ba}, H_{ba}\}$. We denote the threshold of RED$-$VBR mechanism as $D^p = \{L^p_b, H^p_b\}$ in this paper. The inclusion property of three thresholds are $D^h < D^p < D^l$. Initially, $|\beta|$ in Figure 1 is $\rho^e_{max(t+1)}$. β is adjusted according to ρ_t at run time[4]. In contrast, optimistic rate control does not change thresholds, D^l, D^h. $|\alpha_L|$ is $\rho^e_{max(t+1)} - \tau \times \gamma$, and $|\alpha_H|$, $\rho^e_{max(t+1)}$. Buffer can be divided into three areas by D^l,D^h in optimistic rate control: **safe**, **warning** , **emergent**. Safe area corresponds to $[L_{ba}, H_{ba}]$, and warning area either $[L_b, L_{ba})$ or $(H_{ba}, H_b]$. Emergent area is either $[0, L_b)$ or $(H_b, B_s]$. When S_t is in the warning area, optimistic rate control predicts S_{t+1} will be the safe area. In contrast, RED$-$VBR immediately sends λ_t in either $S_t > H^p_b$ or $S_t < L^p_b$.

λ^{RV}_t describes the amount of extra bandwidth required to make $S_{t+1} = (L^p_b + H^p_b)/2$, while, λ^{OR}_t in optimistic rate control algorithm requests the amount of bandwidth to make either $S_{t+1} = L_{ba}$ or $S_{t+1} = H_{ba}$. It is the warning area. Optimistic rate control predict a move into the safe area to be accomplished by the self-stability of flow.

Figure 2 illustrates the algorithm for optimistic rate control. The algorithm corresponds to function f in Eq. (4). E in Figure 2 is a variable to count how many service rounds we have waited for the self-stability of flow. *packet_lost()* tests whether any packet had been lost before or not. It returns 1 when any packet had been lost. Otherwise, 0 is returned

3.3   Critics

Overlapping buffer organizations in Figure 1 generates four sub-areas by D^h,D^p and D^l. In case of the high end of the buffer, the areas are denoted as $A(< H_{ba})$,

```
switch Sₖ{

    case warning :
            if( packet_lost() ){
(1)             send feedback;
(2)             E = 0;
            }else
            if( E == MXC ){
(3)             send feedback;
(4)             E = 0;
            }
            E = E + 1;
            break;

    case emergent :
(5)     send feedback;
(6)     E = 0;
        break;

}
```

Fig. 2. Optimistic rate control algorithm

$B(> H_{ba} \wedge\, < H_b^p)$, $C(> H_b^p \wedge\, < H_b)$, $D(> H_b \wedge 0)$. B area is not recognized in RED−VBR, in contrast, optimistic rate control predicts the self-stability of flow if $S_t \in B$.

We use $pj^M(x)$ to denote the probability of packet loss in x area, and use $ps^M(x)$ to denote the probability of $S_t \in x$. Here, M is either 'O' or 'V' to indicate optimistic rate control mechanism and RED−VBR respectively. Probability that lines (1),(2) in Figure 2 are executed is denoted as $p(MXC)$ and defined in Eq. (8).

$$p(MXC) =\approx \frac{1}{MXC-1} \times (1 - pj^O(B,C))^{MXC-1} \times pj^O(B,C) \qquad (8)$$

$$(pj_1^O(B,C) = pj_2^O(B,C) = ... = pj_x^O(B,C))$$

$p(MXC)$ represents how well we can exploit the self-stability of flow by optimistic rate control. In other words, it is probabilty that $packet\_lost()$ function returns 1 in Figure 2. Figure 3 illustrates the possibile pattern of flow according to $p(MXC)$. (B) in Figure 3, ($p(MXC) \approx 0$), implies that optimistic rate control rarely send λ_t compared to RED−VBR. (A) in Figure 3 shows that more λ_ts are sent in optimistic rate control, however, the burstiness of production rate is less than that of RED−VBR. In respect of the smoothness of traffic, optimistic rate control is superior to RED−VBR as shown in Figure 3. Packet loss might be severe in optimistic rate control in the case of (A) in Figure 3 compared to RED−VBR.

$ps^O(B,C)$ is proportional to $pj^O(B,C)$, since λ_t^{OR} causes $S_{t+1} \in [B,C]$. Higher $pj^O(B,C)$, more λ_t is sent. $ps^O(B,C)$ increases with λ_t, meanwhile, $ps^V(B) \propto \frac{1}{pj^V(B)}$ is ensured between $pj^V(B)$ and $ps^V(B)$, since λ_t causes $S_{t+1} \in A$. Higher $pj^V(B)$, less λ_t^{RV} is sent.

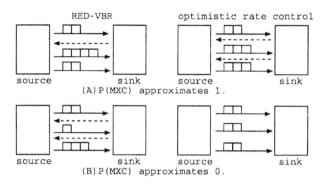

Fig. 3. Possible patterns of flow in accordance woth $p(MXC)$ (Boxes on arrow indicates packet)

4   Performance Evaluation

This section describes the experimental environment and then shows the results of performance evaluation.

4.1   Environment

Linux operating systems were installed into two separate IBM PCs with both Pentium 120MHz MMX and 32 Mbytes main memory. MuSCLE was implemented in both systems. User gives MuSCLE QoS parameters R_b, TAU, RHO, MXC. 100 Mbps hub was used to bridge two PCs, and 10/100 Mbps Intel Eternet ExpressPro was used as network card. 4 clients/servers are spwaned. They communicate via MuSCLE.

We obtained the trace of GOP size from a practical Holliwood movie, *Robocop*. It was encoded in 15 frames per GOP, and all processes in Figure 4 are assumed to send/receive data as much as the size of each GOP. 4 clients consume data as much as recorded in the trace per 0.5 second, while, servers try to send data at the rate of 150Kbytes/sec. Exhaustively investigating about R_b, we found it that the largest of number of self-stability of flow could be exploited when we used 150 Mbits/sec. Used QoS parameters are as follows: $MXC = 4,6,8$, $RHO = 1,2,3$, $TAU = 0.01,0.05,0.1,0.15,0.2$.

MuSCLE sends about 40 Kbytes data every 250 milliseconds for $R_b = 150$ Kbytes/sec. It can be increased up to 60 Kbytes maximally, while can be shrinked down 0 Kbytes according to λ_t.

4.2 $p(MXC)$, Feedbacks and Packet Loss

The performance of optimistic rate control is subject to $p(MXC)$. When $p(MXC)$ is close to 0, optimistic rate control shows better performance rather than

RED$-$VBR. Figure 4 shows the variation of $1 - p(MXC)$ in accordance with
RHO.

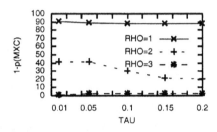

Fig. 4. $1 - p(MXC)$ depending on $RHO,(MXC = 4)$

The number of feedback message varies according to $p(MXC)$ as Figure 3
depicts. Figure 5 shows the variation of feedback messages according to QoS
parameters. When $pj^O(B,C)$ is 0.1, Eq. (8) produces 0.9 approximately. A line
in the upper portion of Figure 4 corresponds to $pj^O(B,C) \approx 0.1$. However, the
rest is not be calculated through Eq. (8) assuming that all $pj^O(B,C)$s are same
at any t-th service round. If a set of four $pj^O(B,C)$s is (0.2,0.2,0.9,1), Eq. (8)
generates about 0.56, and it is illustrated in the second highest line in Figure
4. (0.2,0.2,0.9,1) means that packet loss happens when E counter in Figure 2
reaches 4. If we set $RHO = 3$, the required bandwidth had been quitely under-
estimated. Therefore, the inclusion property of thresholds $,D^l > D^p > D^h$, does
not persist. On the contrary, $D^p > D^l > D^h$,(e.g., $H_b^p < H_{ba} < H_b$) continues
for all sessions. Even though RED$-$VBR mechanism sends λ_t, S_t is recognized
to be within the safe area in optimistic rate control algorithm. Henceforth, lines
(1),(2),(3) and (4) will not be executed in the algorithm depicted in Figure 2.
Both $pj_1^O(B,C)$ and $pj_2^O(B,C)$ are 0.

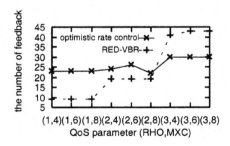

Fig. 5. The number of feedback depending on (RHO,MXC)

Through Figure 5, we can know that a flow would like to be pattern (A) in Figure 3 at $RHO = 3$. Meanwhile, pattern (B) in Figure 3 will be shown at both $RHO = 1$ and $RHO = 2$. Figure 5 shows the number of feedback message according to RHO. In the case of $RHO = 3$, we can find more feedbacks in optimistic rate control rather than RED−VBR.

Figure 6 illustrates how many bytes were lost per a packet in average. It implies that how many resources needs for seamless networking and how many bandwidth are more needed to lossless service.

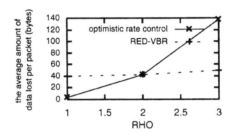

Fig. 6. The average size of data lost $(TAU=0.1)$

As addressed above, more packet loss is shown at $RHO = 3$. At $RHO = 2$, only a few bytes are lost smaller than RED−VBR.

4.3   Traffic and Required Bandwidth

Traffic consists of two sorts of packets which one is affected by feedback and the other is not. Eq. (9) produces the average traffic in both mechanisms.

$$Traff = R_b \times (K - F) + (F_m + PR) \times F \qquad (9)$$

K indicates the total number of service round composing a session. F is the number of feedback in whole session, while, F_m is the size of feedback message. Henceforth, $R_b \times (K - F)$ is to calculate the amount of data which had not been affected by feedback. $(F_m + PR) \times F$ indicates the number of packets which had been affected by feedback.

Figure 7(A) represents how many packets are sent by MuSCLE in both mechanisms:optimistic rate control, RED−VBR. Figure 7(B) illustrates how much bandwidth should be needed in these mechanisms. Required bandwidth is the amount of bandwidth needed to support optimal presentation quality of video. It is computed base on an assumption that packet loss is evenly distributed in whole GOPs composing the video. The bandwidth can be calculated by Eq. (10).

$$RB = Traff + LR \qquad (10)$$

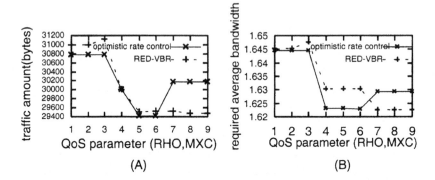

Fig. 7. (A):traffic , (B):required bandwidth for optimal presentation quality

Traff indicates the traffic computed from Eq. (9), and *LR* is the result from Figure 7(B). Both RB_{OR} and RB_{RV} shown in Figure 7(A). We can know that more bandwidth is required when we use optimistic rate control mechanism at $RHO = 3$ by packet loss. Optimistic rate control needs less bandwidth to support the optimal quality of VBR video rather than RED−VBR.

4.4   Smoothness

Smooth traffic can be evaluated by the fluctuation of production rate. It is depicted in Figure 8.

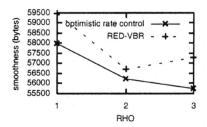

Fig. 8. Smoothness ($TAU=0.1$)

Figure 8 is generated by Eq. (11). RB_{act}^x is an actual bandwidth computed by production rate at the x-th round, and RB_{opt} is the value plotted in Figure 7(B). The difference between them produces the degree of fluctuation. K means by the total number of service round.

$$\frac{1}{K} \times (\sum_{x=1}^{K}(|RB_{act}^x - RB_{opt}|) \times \gamma) \tag{11}$$

We can intuitively predict such a smooth traffic from Figure 3. Even though feedback message request small amount of bandwidth variation, the amount of traffic changed is small rather than RED−VBR.

5   Conclusion and Future Works

We propose a rate control mechanism to support optimal quality of video without modifying encoding method. It is meaningful in that traffic is made smooth without any special modification of media structure. Second, the small amount of bandwidth can also be saved by optimistic rate control. Hundreds of thousands of connections are usually outstanding in the Internet, hence, such a small difference might propage up into remarkable upgrade. Third, optimistic rate control mechanism is supported via MuSCLE, a transport layer for multimedia data including VBR encoded video.

The current version of optimistic rate control mechanism statically deals with thresholds of buffer to signal the state of flow. Adjusting the thresholds to the variation of flow is needed for better adaptivity to practical multimedia networking environment. Also, we should negotiate the average encoding rate of video data which can take full advantage of the self-stability of flow.

References

1. V. P. Kumar, T. V. Larkshman, D. Stiliadis: Beyond Best Effort: Router Architecture for the Differentiated Services of Tomorrow's Internet. http://www.lucent.com/dns/products/ps6400.html
2. James D. Salehi, Zhi-Li Zhang, James F. Kurose, Don Towsley: Supporting Stored Video: Reducing Rate Variability and End-to-End Resource Requirements through Optimal Smoothing. ACM SIGMETRICS (1996).
3. H. Zhang, Edward W. Knightly: RED-VBR : A new approach to support VBR-video on packet switching network. In Proc. of NOSSDAV (1997).
4. Edward W. Knightly, Jingyu Qiu: Measurement-Based Admission control with Aggregate Traffic Envelopes. Proc. of 10th Int. Tyrrhenian Workshop on Digital Communications. (1998).
5. Edward W. Knightly: Resource Allocation for Multimedia Traffic Flows using Rate-Variance Envelopes. ACM Multimedia Systems Journal (1999) (to appear)

Session IV: Mobile Data Replication and Caching

Peer Replication with Selective Control*

David H. Ratner, Peter L. Reiher, Gerald J. Popek, and Richard G. Guy

Computer Science Department
University of California, Los Angeles
{ratner,reiher,popek,rguy}@cs.ucla.edu

Abstract. Many mobile environments require optimistic replication for improved performance and reliability. Peer-to-peer replication strategies provide advantages over traditional client-server models by enabling any-to-any communication. These advantages are especially useful in mobile environments, when communicating with close peers can be cheaper than communicating with a distant server. However, most peer solutions require that all replicas store the entire replication unit. Such strategies are inefficient and expensive, forcing users to store unneeded data and to spend scarce resources maintaining consistency on that data.
We have developed a set of algorithms and controls that implement *selective replication*, the ability to independently replicate individual portions of the large replication unit. We present a description of the algorithms and their implementation, as well as a performance analysis. We argue that these methods permit the practical use of peer optimistic replication.

1 Introduction

Mobile computers frequently experience variable, sporadic, unpredictable, or weak connectivity. Optimistic replication is one well-known method of providing high quality data access in the face of such network conditions [3, 4, 7, 12, 14]. Replication improves performance and reliability by creating multiple copies of important data; optimism allows replicas to be independently updated.

There are two models of optimistic replication. In the *peer-to-peer* (or simply peer) model, all replicas are equals, allowing any replica to communicate with any other. Any-to-any communication is important in mobile situations, when a group of laptops is well-connected within itself but weakly connected or disconnected from distant servers. Additionally, when traveling long distances, often communication with a local partner is easier, more efficient, and more cost effective than synchronization over long-distance links. Finally, any-to-any communication provides advantages for environments where the network topology is variable or unpredictable: replicas are allowed to communicate, exchange updates, and synchronize with whoever is available, rather than waiting for specific servers to become accessible.

* This work was sponsored by the Advanced Research Projects Agency under contract DABT63-94-C-0080.

H.V. Leong et al. (Eds.), MDA'99, LNCS 1748, pp. 169–181, 1999.

However, *client-server* models have traditionally provided superior replication control. Whereas all peers typically store the same set of data, clients in the client-server model can store any subset of the server's data. This flexibility permits clients with less power and storage capacity than servers to still replicate data: clients simply select the desired data set and ignore the remainder of the server's data. Clients can therefore replicate very efficiently as they reduce disk space by storing only the relevant data, communication load by only transferring relevant data, and computational load by only maintaining consistency on the data being used.

Our goal is to combine the replication control of the client-server model with the rich information-exchange capability of the peer model. We call our solution *selective replication*. In the volume scenario,[1] we allow individual files from the volume to be replicated independently. Each volume replica stores just the files from the volume that it deems important. Incorporating selective replication control into the peer model required extensive algorithmic modifications, but we discovered that the flexibility could be added with little or no negative impact, and with substantial benefit. Our solution has been implemented in three replication systems, FICUS [7], RUMOR [2], and ROAM [9], and is the basis for hoarding systems such as SEER [5]. The remainder of the paper will discuss the work in the context of ROAM, as that is the most advanced of the three systems.

2 Optimistic Replication

The guiding principle of optimistic replication is that all accessible replicas should be available for full use, including update. Even isolated or partitioned users can generate updates, which is required for mobile computing. However, partitioned updates implies allowing concurrent updates and resulting *conflicts*, so optimistic schemes must reliably detect these conflicts after the fact. Once detected, conflict resolution must occur before normal file activity on conflicted files resumes. Experience with optimistic replication has shown that conflicts are rare and often automatically resolvable [11].

ROAM [9] provides optimistic replication with peer-based, file-level selective replication. Any file within the volume can physically reside at an arbitrary set of the volume replicas. File replicas can be dynamically added or deleted at any time. ROAM is implemented entirely at the user level, and currently runs on Linux and FreeBSD. ROAM uses a multi-level clustering technique capable of providing hundreds to thousands of peer-based replicas.

ROAM maintains data consistency with a periodic synchronization process called *reconciliation*. Reconciliation is a pairwise process between two replicas: a target replica synchronizes with a source by "pulling" the necessary data and status for shared files. When completed, the target knows all information known at the source. Scalability in ROAM is achieved, in part, by "clustering" the reconciliation process.

---

[1] A volume is smaller than a file system but larger than a directory. For example, a user's home directory and all sub-directories might constitute a volume.

All replicas need not be mutually accessible, but they must form a connected graph. As replicas independently communicate with each other, information is distributed and gossiped to all participants. The particular communication patterns among replicas form the *reconciliation topology*.

Updates are tracked using *version vectors* [8], which is a vector of counters, with one counter per replica. Each counter i tracks the total number of known updates generated by replica R_i. Each replica independently maintains its own version vector for each replicated file; comparing two version vectors compares the update histories of the corresponding file replicas.

Part of maintaining consistency is performing *garbage collection*—the deallocation of resources held by unnamed file system objects [1]. While a relatively simple process in a centralized system, dynamic naming and potentially long-term communication barriers make garbage collection more difficult in partially replicated, distributed systems. ROAM uses a fully distributed, two-phase, coordinator-free algorithm to ensure proper distributed garbage collection. ROAM additionally interlaces this proven-correct algorithm with data-removing optimizations to free disk space and remove user data in the optimal time possible.

3   Selective Replication Design

Instead of a purely file-granularity design, we maintain the volume as a large-granularity abstraction and permit fine-grain file replication within each volume. Volumes provide several benefits, even if replication is required at a finer granularity, including integrity boundaries, locality and performance benefits, and assistance in naming.

But since we require the flexibility of fine-grain replication within the large-grain volume, we therefore provide the ability to selectively replicate individual files within the volume. *Partial volume replicas* only maintain the data structures and replica information for the selected portion of the volume that they store. Figure 1 depicts a partial volume replica example.

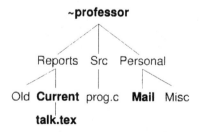

Fig. 1. A partial volume replica. Files listed in bold are stored locally; the others are not.

3.1   Maintaining Replication Information

Sites must determine which files are stored at which replicas to maintain data consistency. Many different solutions are possible. Centralized schemes and replicated schemes based on conservative protocols place undesirable restrictions on when replication factors can change. Therefore, we implemented an optimistic strategy for the maintenance of replication information.

Each replica maintains a *status vector* for each locally stored file. The status vector consists of N elements, where N is the number of volume replicas. Each element is a storage *status value* indicating what information the corresponding volume replica stores about the file.

The two most user-visible status element values (*data* and *nothing*) indicate whether a file's contents are stored at the particular volume replica. Each storage site records its view of where copies of the file reside, and sites that don't store the file (volume replica 2) have no storage overhead. Status vectors are only locally maintained for locally stored files.

Since status vectors are independently and optimistically maintained structures, changes to replication status while partitioned can generate inconsistencies, resulting in *conflicting* status vectors. For example, volume replica 1 could contain the tuple {2, *data*} in its status vector while volume replica 3 has the tuple {2, *nothing*}, resulting in a conflict. For flexibility, we permit any site to order dropping of a replica on any other site, so a site's own status vector is not necessarily correct even for itself.

Conflicting status vectors are resolved by applying a version vector approach to the status vector. Each element is implemented as a monotonically increasing counter, like the version vector. A simple function maps integers to the set of status values; with this approach, two status values can be compared using the standard integer "greater-than" function. Conflicting status vectors can always be automatically resolved by performing a pairwise comparison of status values and selecting the larger one in each case.

3.2   Local Data Availability

The set of all locally stored files at a given volume replica forms a forest of trees. If the forest is permitted to be disconnected, communication with another volume replica would be required to traverse through the intermediate, locally unstored directories that logically connect the disconnected trees. During periods of network partition or disconnection, locally stored data would no longer be available, forcing users to pay close attention to the physical mapping of the tree structure onto the set of volume replicas.

To ensure local availability, we automatically enforce *full backstoring*. Each locally stored file must have its parent directory stored locally as well (the invariant does not apply across volume boundaries). Figure 2 illustrates full backstoring as applied to the partial volume replica from Figure 1.

A common alternate solution employs "prefix pointers" [17], a second directory structure used to connect the user's disconnected namespace. While prefix-pointer solutions avoid storing intermediate directories, they must maintain an

Fig. 2. Full backstoring as applied to the partial volume replica from Figure 1. To store the `talk.tex` and `Mail` files, the intermediate directories must be stored as well. Shaded subtrees are not stored locally.

independent directory structure and integrate it into the user's namespace. Full backstoring avoids the complexity of dual directory structures, and is therefore simpler to implement. Since directories are typically much smaller than files, full backstoring consumes only a small percentage of the available disk space.

UNIX hard links pose problems for full backstoring. Given one file name, it is difficult to find all hard links to the same file without scanning all volume replicas. Therefore, we only guarantee the full backstoring of one link; additional paths can be stored at the user's request using the replication tools discussed in Section 3.3. However, the problem is not serious in practice, because hard links are rare, and many non-UNIX systems do not even support them. Symbolic links, which are far more common, cause no problems for full backstoring.

3.3   Replication Controls

A volume-based replication service has a single policy for file creation—all files are replicated at all volume replicas. The only necessary controls are those that create, add, or delete volume replicas. Since these actions are relatively rare, simple administratively oriented solutions suffice. Robust selective replication implies multiple policies for file creation. Furthermore, users' desires and data demand patterns change, requiring dynamic addition and deletion of file replicas.

Both novices and sophisticated users need reasonable default replication policies. By default, we store new files wherever the parent directory resides, which has proven a reasonable approach. However, since volume roots are always fully replicated, their children, both directories and files, would be created fully replicated as well. Therefore, we provide user-specified replication *masks* on a per-directory basis, allowing users fine control of replication.

Users may also need to dynamically change the replication factors of individual files; that is, *delete* and *add* file replicas. These actions can occur locally or remotely on another replica's behalf.

Replica deletions should only affect storage locations, not the file's existence or physical data. However, data can be inadvertently lost when the replica being deleted contains the most recent version. To guard against data loss, another replica should synchronize the file with the replica being deleted. *Reverse reconciliation* is not always possible, however, as network partitions can make other replicas temporarily inaccessible. We allow users to bypass reverse reconciliation when necessary, although it is an unsafe operation.

Replica deletion at one site is indistinguishable from new object creation at another. Guy [1] describes a solution for resolving the create/delete ambiguity, which we adopt. We keep a temporary record of the action and executes a distributed algorithm to guarantee both that all replicas learn of the action and that the record is eventually removed. For this purpose, the temporary record is a new state, *dropping*. All data-storing replicas participate in a distributed algorithm that guarantees that all replicas correctly learn of the transition. Full details of this algorithm can be found in [10].

Adding file replicas is simpler than deleting them, because there is no risk of data loss, and there is no create/delete ambiguity (by design, the lack of a delete record implies add). The only additional action is possibly adding extra directories, to enforce full backstoring.

4   Maintenance of Consistency

Data consistency in a selectively replicated environment requires more complete algorithms and stronger communication assumptions than in a fully replicated environment. Without selective replication, a given replica A can reconcile all data and learn about all updates by communicating with only a single other site B. In a selectively replicated environment, such as depicted in in Figure 3, A may store a different set of data from B, as shown in the figure. If A's set is not a subset of B's, then A can neither reconcile all of its data with B nor learn about all new updates from B. Reconciliation with other sites is required—site D in the case of Figure 3. Forcing B to maintain enough information such that A could completely synchronize with it would be tantamount to requiring that B store a superset of A's data, nullifying most if not all of the advantages of selective replication.

Consistency in this context requires the distribution of updates and the garbage collection of unnameable file system objects.

4.1   Distribution of Updates

Reconciliation maintains data consistency, and the *reconciliation topology* is the communication pattern used between replicas. One replica *pulls* data from another, and in doing so learns all relevant knowledge stored at the source. While a two-way protocol has its advantages, notably performance, a pull-only strategy is more general.

The underlying algorithms are topology-independent: they only require that information can flow from any replica to any other in a finite number of steps and through a finite number of other replicas. However, different topology patterns

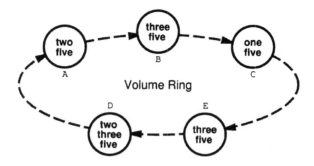

Fig. 3. An example of four selectively replicated files, named by their replication factor. Circles represent volume replicas; arrows indicate data flow. In this topology, reconciliation cannot make progress without forcing the non-data-storing replicas to store file data.

yield different results in terms of performance and message volume complexity [16]. ROAM uses a ring topology for propagating updates, with enhancements to handle selective replication. Also, ROAM's scalability is based on clustering enhancements to the ring topology [9].

In a ring topology, participants reconcile with the "next" member in the set of volume replicas V, ordered by replica identifier. The ring is dynamically and independently reconfigured by each replica when the set V changes in size. Also, if the preferred reconciliation partner is inaccessible, a node reconciles with the next accessible replica. The clustering enhancements [9] group replicas into local collections to manage the problems caused by mobility, such as "stretching" or "contorting" the ring of replicas. In this way, we provide good performance and efficiency even when mobile replicas effectively destroy the previous ring.

The *adaptive ring* requires only a linear message complexity of $O(|V|)$ messages to propagate information to all replicas. Replicas gossip, transmitting information received about the state of other replicas to those which follow. Thus, the topology does not require point-to-point links inter-connecting all replicas.

Unfortunately, the adaptive ring as described above only operates correctly in a fully-replicated scenario. As seen in Figure 3, selective replication invalidates the implicit assumption that any node can receive and propagate all information via any other node.

A *multi-ring* topology retains the ring's robustness but still correctly supports selective replication. The multi-ring strategy associates an adaptive ring with each individual file. The file ring is also adaptive.

Reconciliation for each file proceeds along that file's own per-file ring, as illustrated in Figure 4. However, we must address the performance implications raised by a multi-ring strategy. In practice, a ring per file is impractical. ROAM consolidates files in a volume into *replica sets*, groups of files that are all stored at common volume replicas. Replica set information is managed and maintained in an optimistic fashion, like the data itself, based on replica-local information. As part of reconciliation, the local replica will adjust not only versions of local data but also meta-data, such as where the other replicas are physically stored. Full details on how replica sets are formed and used can be found in [10].

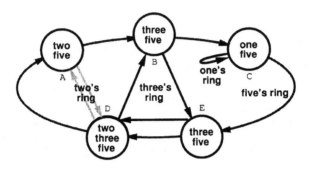

Fig. 4. The same four files as in Figure 3, with each per-file ring indicated. Arrows indicate the data flow between volume rings, labeled by the data transported over them. Together all four rings comprise the multi-ring topology for the given replication pattern.

4.2 Garbage Collection

Garbage collection deallocates file system resources (disk space, inodes, etc.) after file removal. Actual deallocation is straightforward, but determining the correct time to do so is difficult. Dynamic naming[2] and long-term communication delays imply that files might temporarily have no local links, but still be globally accessible.

5   Performance Evaluation

This section describes the measurements, experiments, and benchmarks used to evaluate selective replication, as well as the results and conclusions that can be drawn from them. We begin with a description of the data overhead imposed by selective replication, describe the performance of reconciliation, and conclude with some comments on real use and future evaluation work.

The measurements and benchmarks were performed on ROAM.

---

[2] Files in systems like UNIX can have multiple names created dynamically, either by link or rename operations.

5.1   Data Overhead

Selective replication requires adding a status vector to the data structures already stored for full replication. Each replica in the *data, attributes-only*, or *dropping* state stores a status vector consuming 4N bytes, where N is the number of volume replicas. Replicas in the *nothing* state store no attributes at all.

We studied the effect of selective replication on the disk overhead using five different replication patterns. The volume used is a 13.6MB volume, consisting of 307 files, of which 38 are directories. It contains large files (up to 3.5MB) and small files (a few bytes), large directories (the volume root contains 151 files) and small directories. Using the selective replication tools, we created five different replication "patterns," summarized in Table 1. For ease of recognition, we have given each pattern a simple, mnemonic name.

Table 1. The five replication storage patterns used for studying the effects of selective replication. Each pattern has a different subset of the 13.6MB volume. The complete volume stores 307 files, of which 38 are directories. In each case, the total number of file system objects *includes* the number of directories.

Pattern	Objects	Directories	Size
small	173	38	2.6MB
space-saver	295	38	5.75MB
inode-saver	218	38	11.21MB
subtree	256	31	12.9MB
full	307	38	13.6MB

We created two replicas, performing the selective replication changes and measuring the results at the second replica. Replica 1 stored a fully replicated volume; replica 2 stored different subsets of the volume, according to the five replication patterns. Figure 5 illustrates the results, showing two different curves. The first curve indicates the percentage of disk overhead as a function of the amount of locally stored user data, which is one measure of the cost of selective replication. The second shows the savings in disk space provided by selective replication (including the overhead), which is one benefit of selective replication.

Replication pattern "small" has the smallest average file size; we therefore expect it to have the highest percentage of overhead. As the average file size gets larger, the percentage of overhead becomes smaller.

The second set of bars in the graph illustrates a benefit of selective replication. We preserve access to all files in the volume, while potentially saving the user up to 80% of the required disk space, as in the case of pattern "small."

5.2   Synchronization Performance

The key performance figure for ROAM is the actual performance of synchronization under different selective replication patterns and workloads. We performed

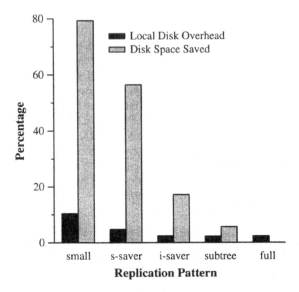

Fig. 5. Selective replication disk overhead for the 13.6MB volume. The replication patterns are described in Table 1.

our experiments with two portable machines connected by a 10Mb Ethernet and with minimum daemons running. The first machine is a Dell Latitude XP with a 486DX4 running at 100Mhz and with 36MB of main memory. The second is a TI TravelMate 6030 with a 133Mhz Pentium and 64MB of main memory. Reconciliation was always performed transferring data from the Dell machine to the TI machine. In other words, the reconciliation process always executed on the TI machine.

We utilized the previous five replication patterns at the local (TI) machine. Additionally, since reconciliation performance depends heavily on the sizes of the files that have been updated, we varied the amount of data updated at the remote site (and therefore the amount to be transferred by reconciliation to the local site) from 0 to 50%. Data files are selected at random for update; they are not predetermined. We performed seven runs for each data point; the results are illustrated in Figure 6.

Selective replication saves the user up to half of the cost of replication. The amount of data updated at the remote side in pattern "small" makes little difference, since almost none of it is replicated at the local replica. In contrast, for the "full" replication pattern, reconciliation time rises by almost one-third when the amount of data being modified increased from 0 to 50%. The figure demonstrates a second clear benefit to selective replication, in addition to the disk space savings demonstrated above: reconciliation takes less time.

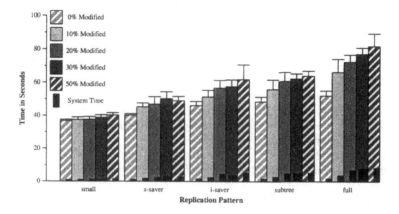

Fig. 6. Reconciliation performance for the 13.6MB volume. The replication patterns are described in Table 1. 95% confidence intervals are shown. Except where indicated, all measurements are of elapsed time.

Even when nothing was modified at the remote site, reconciliation took longer under replication pattern "full" than replication pattern "small." Since more files are stored locally in replication pattern "full," reconciliation spends more time detecting if any updates have been generated locally (to update its local data structures and version vectors), regardless of whether or not updates were applied at the remote site.

Selective replication has been used in production with FICUS, RUMOR, and ROAM. Usually, use was confined to a static model: users dropped specific files or sub-trees from specific replicas, and didn't again change the replication patterns. Typically, such uses were motivated by space and/or time restrictions. Additionally, files that didn't need to be stored at certain replicas were dropped from them, to reduce reconciliation time.

6   Related Work

CODA [4, 12] is an optimistic replicated file system constructed on the client-server model, as opposed to the peer architecture. CODA trivially provides selective-replication control at the clients, but not at the replicated servers which are, by definition, peers. Additionally, the clients cannot inter-communicate due to the inherent restrictions of the client-server model. The restrictions dramatically simplify the algorithms needed to manage consistency, at the cost of limiting the system's utility for mobile workgroups and other environments.

CODA has greatly optimized the communications and synchronization between client and server, especially in environments with weak connectivity [4, 6]. Additionally, CODA uses a vastly simplified garbage collection strategy at their replicated servers, based on logging directory operations. The viability of

CODA's solution depends on how often the logs become full, as dropping some log information can lead to unnecessary conflicts. Log wrap-around could result from long disconnections or a plethora of file system activity. Should the CODA strategy be applied in a mobile environment, where disconnections are commonplace, the log wrap-around strategy could result in a large number of unnecessary conflicts.

The BAYOU system [14] is also a replicated storage system. Like FICUS and RUMOR, BAYOU is based on the peer-to-peer architecture. However, BAYOU does not yet support any form of selective replication.

The DECEIT file system [13], places all files into one "volume" and allows each individual file to be replicated independently with varying numbers of replicas. However, DECEIT employs a conservative writer-token approach to replication, and thus cannot provide the high availability offered by optimistic mechanisms. Also, DECEIT cannot tolerate long-term network partitions.

The university version of the LOCUS operating system [15] provides volumes and allows selective replication within the volume. However, the approach taken toward replication is not strictly optimistic.

7 Conclusions

Peer models are necessary in a wide variety of environments, but have traditionally provided less flexibility than client-server models. Selective replication provides the replication flexibility of the client-server model while maintaining the advantages of a peer architecture. By enabling any-to-any communication and providing the ability to store dynamic subsets of replication units, the functionality and flexibility of the two models have effectively been merged.

Even without selective control, peer replication is important for the rich information model it provides. However, there are many environments where peer solutions are simply unusable in practice without selective replication due to the inherent costs of traditional peer models. Selective replication allows peer solutions to be effectively and efficiently utilized in scenarios that otherwise would prove impractical. With the rapid rise of mobility and mobile workgroups, peer solutions will be required even more than before, making selective replication an integral part of any replication service designed for a mobile world.

References

1. R. G. Guy, G. J. Popek, and T. W. Page, Jr. Consistency algorithms for optimistic replication. In *Proceedings of the First IEEE International Conference on Network Protocols*. IEEE, Oct. 1993.
2. R. G. Guy, P. L. Reiher, D. H. Ratner, M. F. Gunter, W. Ma, and G. J. Popek. *Rumor*: Mobile data access through optimistic peer-to-peer replication. In *Advances in Database Technologies:ER '98 Workshops on Data Warehousing and Data Mining, Mobile Data Access, and Collaborative Work Support and Spatio-Temporal Data ManagementProceedings of the*, number 1552 in Lecture Notes in Computer Science, pages 254–265. Springer Verlag, Nov. 1998.

3. J. S. Heidemann, T. W. Page, Jr., R. G. Guy, and G. J. Popek. Primarily disconnected operation: Experiences with Ficus. In *Proceedings of the Second Workshop on Management of Replicated Data*, pages 2–5. University of California, Los Angeles, IEEE, Nov. 1992.

4. J. J. Kistler and M. Satyanarayanan. Disconnected operation in the Coda file system. *ACM Transactions on Computer Systems*, 10(1):3–25, 1992.

5. G. H. Kuenning and G. J. Popek. Automated hoarding for mobile computers. In *Proceedings of the 16th Symposium on Operating Systems Principles*, pages 264–275, St. Malo, France, Oct. 1997. ACM.

6. L. B. Mummert, M. R. Ebling, and M. Satyanarayanan. Exploiting weak connectivity for mobile file access. In *Proceedings of the 15th Symposium on Operating Systems Principles*, pages 143–155, Copper Mountain Resort, Colorado, Dec. 1995. ACM.

7. T. W. Page, R. G. Guy, J. S. Heidemann, D. Ratner, P. Reiher, A. Goel, G. H. Kuenning, and G. J. Popek. Perspectives on optimistically replicated peer-to-peer filing. *Software—Practice and Experience*, 28(2):155–180, Feb. 1998.

8. D. S. Parker, Jr., G. Popek, G. Rudisin, A. Stoughton, B. J. Walker, E. Walton, J. M. Chow, D. Edwards, S. Kiser, and C. Kline. Detection of mutual inconsistency in distributed systems. *IEEE Transactions on Software Engineering*, 9(3):240–247, May 1983.

9. D. Ratner, P. Reiher, and G. J. Popek. Roam: A scalable replication system for mobile computing. In *MDDS'99: 2nd International Workshop on Mobility In Databases and Distributed Systems @ DEXA'99*, Sept. 1999.

10. D. H. Ratner. Selective replication: Fine-grain control of replicated files. Master's thesis, University of California, Los Angeles, Mar. 1995. Available as UCLA technical report CSD-950007.

11. P. Reiher, J. S. Heidemann, D. Ratner, G. Skinner, and G. J. Popek. Resolving file conflicts in the Ficus file system. In *USENIX Conference Proceedings*, pages 183–195, Boston, MA, June 1994. USENIX.

12. M. Satyanarayanan, J. J. Kistler, P. Kumar, M. E. Okasaki, E. H. Siegel, and D. C. Steere. Coda: A highly available file system for a distributed workstation environment. *IEEE Transactions on Computers*, 39(4):447–459, Apr. 1990.

13. A. Siegel, K. Birman, and K. Marzullo. Deceit: A flexible distributed file system. In *USENIX Conference Proceedings*, pages 51–61. USENIX, June 1990.

14. D. B. Terry, M. M. Theimer, K. Petersen, A. J. Demers, M. J. Spreitzer, and C. H. Hauser. Managing update conflicts in Bayou, a weakly connected replicated storage system. In *Proceedings of the 15th Symposium on Operating Systems Principles*, pages 172–183, Copper Mountain Resort, Colorado, Dec. 1995. ACM.

15. B. Walker, G. Popek, R. English, C. Kline, and G. Thiel. The LOCUS distributed operating system. In *Proceedings of the Ninth Symposium on Operating Systems Principles*, pages 49–70. ACM, Oct. 1983.

16. A.-I. Wang, P. Reiher, and R. Bagrodia. A simulation evaluation of optimistic replicated filing in mobile environments. In *Proceedings of the 18thProceedings of the 18th IEEE International Performance, Computing, and Communications Conference*, Feb. 1999.

17. B. Welch and J. Ousterhout. Prefix tables: A simple mechanism for locating files in a distributed system. *Sixth International Conference on Distributed Computing Systems*, pages 184–189, May 19-23, 1986.

Cache Coherency in Location-Dependent Information Services for Mobile Environment

Jianliang Xu, Xueyan Tang, Dik Lun Lee, and Qinglong Hu *

University of Science and Technology, Clear Water Bay, Hong Kong
{xujl, tangxy, dlee, qinglong}@cs.ust.hk

Abstract. Caching frequently accessed data at the client is an attractive technique for improving access time. In a mobile computing environment, client location becomes a piece of changing information. As such, location-dependent data may become obsolete due to data updates or client movements. Most of the previous work investigated cache invalidation issues related to data updates only, whereas few considered data inconsistency caused by client movements. This paper separates location-dependent data invalidation from traditional cache invalidation. For location-dependent invalidation, several approaches are proposed and their performance is studied by a set of simulation experiments. The results show that the proposed methods substantially outperform the *NSI* scheme which drops the cache contents entirely when hand-off.

1 Introduction

Mobility opens up new classes of applications in a mobile computing environment. Location-dependent information service is one of these applications which are gaining increasing attention [5, 1, 9]. Through location-dependent information service, mobile clients can access location sensitive information, such as traffic report, hotel information and weather broadcasting, etc. The Advanced Traveler Information Systems(*ATIS*) project has explored this in depth [11]. With recent development in micro-cell/pico-cell systems, precise location-dependent information is available, making applications which require precise location data (e.g., navigation maps) a reality.

Caching is a widely used technique in mobile computing environments to improve performance [3, 2, 7]. However, frequent client disconnections and movements across cells make cache invalidation a challenging issue [2]. A lot of work has been done on cache invalidation [3, 8, 6, 4, 12]. Most of the previous work studied the cache consistency problem incurred by data updating (hereafter called *temporal-dependent update*). In a mobile computing environment, besides the temporal-dependent updates, cache inconsistency can also be caused by location changing (hereafter called *location-dependent update*). Location-dependent

---

* The author is now with Department of Computer Science, University of Waterloo, Waterloo, Ontario, Canada.

H.V. Leong et al. (Eds.), MDA'99, LNCS 1748, pp. 182–193, 1999.
© Springer-Verlag Berlin Heidelberg 1999

updates refer to the case where the cached data become obsolete when the mobile client moves. Therefore, cache invalidation should be performed for both temporal-dependent and location-dependent updates. Temporal-dependent and location-dependent invalidations affect each other. For example, if the temporal-dependent update rate is much higher than the location-dependent update rate, the temporal-dependent updates will dominate cache invalidation, and vice versa.

This paper investigates the integration of temporal-dependent and location-dependent invalidation. For location-dependent invalidation, we propose Bit Vector with Compression (BVC), Grouped Bit Vector with Compression ($GBVC$), and Implicit Scope Information (ISI) methods. For temporal-dependent cache invalidation, we adopt the AAW_AT (Adaptive Invalidation Report with Adjusting Window) method [6]. Experiments are conducted under various system parameter settings. When both query throughput and uplink cost are considered, all the three location-dependent invalidation strategies outperform the No Scope Information (NSI) method which drops the cache contents entirely when hand-off. In particular, ISI and $GBVC$ are close to the optimal strategy and better than BVC. The $GBVC$ scheme is the best for most system workloads.

The rest of this paper is organized as follows. Section 2 describes some assumptions in this study. Section 3 proposes three location-dependent invalidation schemes. The simulation model and experimental results are presented in Section 4 and 5, respectively. Finally, Section 6 concludes the paper.

2 Assumptions

The geographical coverage area for the information service is partitioned into *Service Areas*, and each service area is attached with a data server. The service area may cover one or multiple cells. Each service area is associated with an *ID* (SID) for identification purposes. The SID is broadcast periodically to all the mobile clients (MCs) in that service area.

The database associated with each service area is a collection of data items. The *valid scope* of an item value is defined as the set of service areas where the item value is valid. All valid scopes of an item form its *scope distribution*. A scope distribution may be shared by several data items. In a large-scale information system, the number of scope distributions can be very large. Every data server keeps a complete copy of the database, i.e., the same data items are replicated on all the data servers but probably with different values in different data servers. Data servers or $MSSs$ are assumed to know the scope of each data item. When a mobile client moves into a new service area, validity checking of cached data needs to be carried out. When a data item is updated, there is a certain delay, δ, involved to maintain consistency for replicas. For simplicity, we assume that $\delta = 0$. Thus, we do not need to consider the consistency problem caused by update delays.

3   Cache Invalidation Methods

In general, there are two types of cache invalidation strategies, namely *client-initiated method* and *server-initiated method*. In the client-initiated method, the client monitors the states of the cached items and initiates the validity checking procedure. In the server-initiated method, the server monitors the states of the cached items and informs the client to purge the obsolete data. Since temporal-dependent invalidation is usually carried out at the server side, it is difficult for the client to know whether cached data are valid or not. As such, the server-initiated method is usually adopted. In contrast, location-dependent invalidation is caused by the movement of the client. Therefore, it is hard for the server to know the state of the client cached data and the client-initiated method is a simpler approach.

3.1   Temporal-Dependent Invalidation

For temporal-dependent invalidation, periodic broadcasting of invalidation reports at the server side is an efficient strategy [3, 8]. Every mobile client, if active, listens to the reports and invalidates its cache accordingly. In the various invalidation report approaches, adaptive cache invalidation algorithms work well under various system workloads [6]. In adaptive cache invalidation algorithms, the next invalidation report (IR) is dynamically decided based on the system workload.

3.2   Location-Dependent Invalidation

As discussed above, temporal-dependent cache invalidation is initiated by the data server. On the other hand, location-dependent cache invalidation is client-initiated. In order to validate the cached data, the mobile client can send the IDs of the cached items uplink to the data server. The data server decides whether these items are still valid according to the mobile client's current location and sends the checking result back to the client. This method is simple. However, it increases network traffic substantially due to the expensive send-up-checking. To achieve better performance, it is better to attach scope information to the data. In that case, invalidation efficiency depends heavily on the organization of the scope information. In the following paragraphs, we propose several efficient methods to build scope information.

Bit Vector with Compression (BVC)   Recall that every service area is associated with an SID to distinguish it from the others. BVC uses a bit vector to record scope information. The length of the bit vector is equal to the number of service areas in the system and every cached data item is associated with a bit vector. A "1" in the nth bit vector indicates that the data item value is valid in the nth service area while "0" means it is invalid in the nth service area.

For example, if there are 12 service areas in the system, then a bit vector with 12 bits is constructed for each cached data item. If the value of the bit vector for a data item is 000000111000, it means this data item value is valid in the 7th, 8th, and 9th service areas only.

Obviously, the overhead still would be significant when the system is large. Noticing the locality of validity scope, it is possible to perform compression on the bit vector.

Grouped Bit Vector with Compression ($GBVC$) To remedy the significant overhead in BVC, we can build a vector corresponding to only part of the service areas. The reason is two-fold. Firstly, the mobile client seldom moves to a service area that is far away from its home service area. Consequently, partial data scope information is sufficient. Secondly, even if the mobile client moves to a distant service area, it takes quite some time for the client to do so. During this period, the data item may have already been updated on the data server, and therefore the complete scope information about distant service areas is useless. Hence, it is more attractive to store the data item scope information of only adjacent service areas. This is where the idea of Grouped Bit Vector with Compression scheme ($GBVC$) comes from.

The whole geographical area is divided into disjoint districts and all the data service areas within a district form a group. Here, the service area ID (SID) consists of two parts: *group ID* and *intra-group ID*. Scope information attached to a data item includes the group ID and a bit vector (BV) which corresponds to all the service areas within the group. Thus, an attached bit vector has the following form: (group ID, BV).

With the same example used in the previous section, the whole geographical area is further assumed to be divided into two groups, such that service areas 1-6 form group 0 and the rest form group 1. With the $GBVC$ method, one bit is used to construct the group ID and a six-bit vector is used to record the service areas in each group. For the data item mentioned earlier, in group 0, the attached bit vector is $(0, 000000)$; in group 1, the attached bit vector is $(1, 111000)$. As can be seen, the overhead for scope information is reduced in the $GBVC$ method.

When the mobile client checks the data item's validity, it compares the group ID of the current service area with the one associated with the cached data item. If they are not the same, the data item will be invalidated. Otherwise, the client checks the *intra-group ID*th bit in the bit vector to determine whether the cached item should be invalidated.

Implicit Scope Information (ISI) This strategy divides the database into multiple logical sections. Data items with the same valid scope distribution are placed in the same section. Hence, data in the same section share the same location validation information. Different logical sections imply different scope information.

Each possible valid scope of the data item value is specified by a 2-tuple (SDN, SN), where SDN is the scope distribution number and SN denotes the

SID	1	2	3	4	5	6	7	8	9	10	11	12
(SDN)Scope Distribution #1	1	2	3	4	5	6	7	8	9	10	11	12
(SDN)Scope Distribution #2	1			2			3			4		
(SDN)Scope Distribution #3	1		2			3		4			5	

Fig. 1. An Example with Different Distributions

scope number within this distribution. The 2-tuple is attached to the data item during delivery. The cached item i has the format $\{D_i, SDN_i, SN_i\}$, where D_i is the item value. For example, suppose there are three types of scope distributions (see Figure 1) and data item 4 has distribution 3. If item 4 is cached from service area 6 (i.e., $SID = 6$), then $SDN_4 = 3$ and $SN_4 = 3$. That implies that the cached item 4's value is valid in the service areas 6 and 7 only.

The location-dependent report consists of the valid SNs for each scope distribution and the SID of that service area. For example, in service area 8, the MSS broadcasts $\{8, 3, 4, 8\}$ to the clients, where the first three numbers are the SN values for the three scope distributions and the last number is the SID of the current service area. In this way, scope information delivered along with the data item can be greatly reduced.

After retrieving this location-dependent report, for item i, the client compares the cached item's SN_i with the SDN_ith SN in the location-dependent report received. If they are the same, the cached item value is valid. Otherwise, the data item value is invalid. For example, in service area 8, the client checks for the cached data item 4 whose $SDN_4 = 3$ and $SN_4 = 3$. In the broadcast report, the SDN_4th (i.e. 3rd) SN equals to 4. Therefore, the client knows that data item 4's value is invalid.

4   Simulation Model

This section describes the simulation model used to evaluate the performance of the cache invalidation methods proposed in the previous sections. The performance metrics used are system throughput (i.e., number of queries answered per interval per cell) and average uplink cost for answering a query.

4.1   System Model

The system consists of $CellNumber$ (i.e. $CellNumber_x \times CellNumber_y$) cells with hexagonal shapes and $UserNumber$ users. In the simulation model, every cell is assumed to be a service area. Therefore, client moving across service areas is the same as hand-off. Initially, the users are uniformly distributed in all $CellNumber$ cells, i.e. each cell has $UserNumber/CellNumber$ users. During the experiment, every user moves between cells independently according to the same movement pattern.

An MSS serves as a data server for each cell. The database of a data server contains $DatabaseSize$ items and is uniformly divided into $Scope\_N$ classes. The data items in one class have the same valid scope distribution. In other words, there are $Scope\_N$ valid scope distributions in a database. The mean size of a data item's valid scope is $ScopeSize_i(i = 1, 2, \ldots, Scope\_N)$.

The network is modeled as a *preempt resume* server with invalidation report and location-dependent report having the highest broadcast priority and the rest of the messages having the same priority. All other messages are served on an $FCFS$ basis. The bandwidth is always fully utilized so that the throughput represents the true efficiency of the algorithm. Table 1 shows the default system parameter settings used in the experiment.

Table 1. The default system parameter settings.

Parameter	Setting	Parameter	Setting
Simulation Time	200000 seconds	CellNumber	4096(64x64) cells
User Density	20 mobile clients/cell	Database Size	1000 data items
Data Item Size	1024 bytes	ScopeClass	10
Broadcast Period	20.0 seconds	Control Message Size	64 bytes
Downlink Bandwidth	1000 bps	Mean ScopeSize for	1, 4, 4, 16, 16, 64,
Uplink Bandwidth	1000 bps	Each ScopeClass	64, 256, 1024, 4096

4.2   Client Execution Model

Each client is simulated by a process and runs a continuous loop that generates streams of queries and hand-offs independently. The queries are read-only and are location-dependent in the sense that the query results reflect the current location (cell) of the mobile user. Each query is separated from the completion of the previous query by either an exponentially distributed *Think Time* or an exponentially distributed *Disconnection Time*. In the model, each client has a probability p to enter the disconnection mode at the beginning of each broadcast interval. If the desired data items have a valid copy in the client, then an immediate reply is returned. Otherwise, the client sends the data IDs uplink to the server. The server returns the data by broadcasting them on the channel. Once the requested data items arrive, the client retrieves them off the channel and stores them in the cache. The cache page replacement policy is the LRU scheme.

The time that a mobile client spends in a cell follows an exponential distribution with mean time of *Residence Time*. After the client stays in a cell for a while, it moves to another neighboring cell. The destination cell is chosen based on uniform probability (i.e. 1/6 probability for each of the six neighbors). All mobile clients follow the same move pattern independently.

The client query access pattern follows hot-spot and cold-spot with the setting of 80/20, that is, 80% queries access first 100 hot data items and 20% queries access the rest of the database. In both cases, queries are uniformly distributed in the different scope classes. In reality, the user may access a hint of data items whose values are "correct" for another location (cell). For simplicity, it is assumed that no such queries exist and all accesses are directed to the local server. Table 2 shows the default parameter settings for mobile clients.

Table 2. Default parameter settings for mobile clients.

Parameter	Setting	Parameter	Setting
Client Cache Size	10% of database size	Mean Think Time	50.00 seconds
Mean Residence Time	600.00 seconds	Mean Disconnection Time	400.00 seconds
Mean Data Items Ref. by a Query	10 data items	Prob. of Client Disc. per Interval	0.1

4.3   Server Execution Model

The server is simulated by a process which generates a stream of updates with an exponentially distributed *Update Inter-arrival Time*. Unlike queries, all updates are assumed to be randomly distributed across the whole database. For data items with several "correct" values in different cells, the updates arrive at those data servers at the same time. Thus, the updates happen simultaneously for these data items. Furthermore, the server broadcasts invalidation reports and location-dependent reports, which consists of SID, SNs etc., periodically to the clients. The default parameters used to describe the server model are given in Table 3.

Table 3. Default parameter settings for data servers.

Parameter	Setting
Mean Update Inter-arrive Time	500.00 seconds
Mean Data Items Updated Each Time	5 data items

5   Experiment Results

The simulation is carried out using $CSIM$ [10]. We collect data from the first 64th (i.e. 8×8) cells after the system becomes stable (after 100000 seconds in the simulation). Various experimental results are presented in this section. Besides the strategies proposed in the previous sections, two additional strategies, namely No Scope Information (NSI) and *Optimal*, are also included for comparison. NSI represents the strategy in which no extra mechanism is used for location-dependent cache invalidation, i.e., the client simply drops the entire cache after hand-off. *Optimal* denotes an ideal strategy in which the client has complete valid scope information. We evaluate the $GBVC$ strategy for various group sizes and find that 64 is a reasonable choice under our experiment settings. Therefore, we only represent the data collected from the $GBVC$ strategy with group size 64 in this section. For the BVC strategy, the system performance is evaluated under compression ratio 1:1 and 1:2.5, denoted by BVC and $BVC0.4$ respectively.

All of the above strategies are evaluated under the NC scheme for validity checking time, where after hand-off the MC does not make cache invalidation until the query is issued. Both temporal-dependent and location-dependent cache invalidation are realized in the simulation. For temporal-dependent cache invalidation, AAW_AT, which shows the best performance in the absence of location-dependent invalidation [6], is adopted.

5.1 Performance for Changing Mean Residence Time

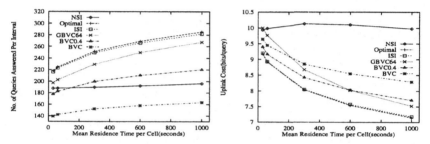

Fig. 2. Performance for Changing Mean Residence Time

System performance is first investigated by varying mean client residence time in a cell from 30 seconds to 1000 seconds. The shorter the duration time in a cell, the more frequent the client hand-off. As shown in Figure 2, ISI and $GBVC64$ improve the throughput and uplink cost substantially over NSI. In particular, ISI performs almost the same as $Optimal$. Although BVC makes use of the additional valid scope information and achieves less uplink cost (due to a higher cache hit ratio[1]) than NSI, its system throughput is worse than NSI. The reason is that BVC introduces too much overhead on the downlink channels and cache storage. On the other hand, due to less overhead, $BVC0.4$ shows better performance than NSI when hand-off occurs less frequently.

All the strategies except NSI are sensitive to client hand-off. When hand-off occurs frequently, the client can easily move out of the valid scope of its cached items. Thus, the effectiveness of location-dependent cache invalidation decreases. When hand-off becomes less frequent (i.e., duration is greater than 600s), all the curves become flat. The reason is that, in this case, location-dependent invalidation seldom occurs in the mobile client and, therefore temporal-dependent invalidation dominates cache invalidation.

---

[1] We do not plot the cache hit ratio due to the space limitation. It is observed that the uplink cost is proportional to the cache miss ratio.

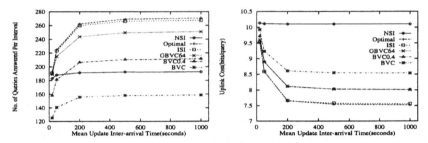

Fig. 3. Performance for Changing Mean Data Update Rate

5.2 Performance for Changing Mean Data Update Rate

Figure 3 presents the results when data update inter-arrival time is varied. Similar to the previous experiments, ISI has the best performance in terms of both throughput and uplink cost and $GBVC64$ is second followed by $BVC0.4$. When updates become intensive (left part of the figures), temporal-dependent invalidation occurs more frequently. Thus, even if the client preserves valid items in the new cell after hand-off, these item values will be purged soon due to frequent temporal-dependent invalidation. Location-dependent invalidation in this case is not a dominant factor in cache invalidation. Hence $GBVC$ and ISI have a similar performance as NSI when the data update rate is high. On the other hand, as the data update rate decreases, $GBVC$ and ISI improve the performance greatly and all the curves become flat. The reason is that the data item values kept after hand-off become more useful and the effect of location-dependent cache invalidation is much more significant than temporal-dependent invalidation.

5.3 Impact of Data Update Rate and Hand-off Rate

To view the effect of relative intensity between temporal-dependent update and location-dependent update, we vary the ratio of mean data update rate to hand-off rate from 0.125 to 16 as shown in Figure 4. When the data update rate is lower

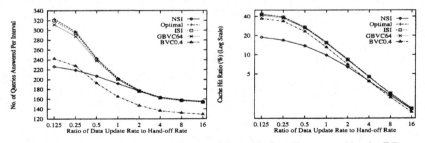

Fig. 4. Performance for Changing Ratio of Data Update Rate over Hand-off Rate

than the hand-off rate, i.e., the ratio is less than one, except for the $BVC0.4$ method, all location-dependent cache invalidation schemes outperform the NSI approach. Due to the high overhead, the performance of the $BVC0.4$ method is worse than the NSI approach when the ratio is greater than 0.3. When the ratio is greater than one, i.e., data update rate becomes high, the location-dependent cache invalidation schemes lose their advantages over the NSI approach. This suggests that location-dependent cache invalidation is desired only when the mean data update rate is low compared with the hand-off rate.

5.4 Impact of System Capacity

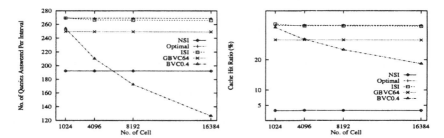

Fig. 5. Performance for Changing Cell Number

In this subsection, the scalability of the location-dependent strategies is e-valuated by varying the cell number from $1024(32 \times 32)$ to $16384(128 \times 128)$ (see Figure 5). The performance of $BVC0.4$ deteriorates rapidly as the number of cells increases, whereas the other schemes are hardly affected. It is interesting to notice that although the cache hit ratio of $BVC0.4$ is higher than that of NSI when there are 16384 cells in the system, it still performs worse than NSI in terms of throughput. This is due to the high overhead of $BVC0.4$.

5.5 Impact of Changing Scope Distribution Number

For a large-scale information system, the scope distribution number might be very large. The adaptability of different location-dependent strategies to the scope distribution number is measured in this set of experiments (refer to Figure 6).

The performance of the ISI method depends heavily on the number of s-cope distributions. When the scope distribution number increases, the system throughput decreases dramatically. On the contrary, all the other schemes remain the same as they do not rely on this system parameter.

In summary, the proposed location-dependent schemes are able to enhance system performance significantly. While the BVC method cannot scale up to system capacity and the ISI scheme does not perform well in a system with a large number of scope distributions, the $GBVC$ method shows the best stability and scalability.

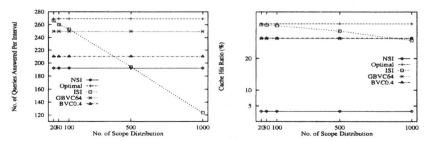

Fig. 6. Performance for Changing Scope Distribution Number

6   Conclusion

In this paper, we have studied cache coherency issues in the context of location-dependent information services and proposed several cache invalidation schemes. Cache invalidation issue caused by both temporal- and location-dependent updates has been addressed. For location-dependent updates, several cache invalidation schemes have been proposed. They differ from each other in scope information organization.

We have conducted a set of experiments to evaluate these schemes in the scenario where temporal-dependent and location-dependent updates coexist. It is found that the proposed strategies are able to achieve a much better performance than the baseline scheme which drops the cache contents entirely when hand-off. While *BVC* cannot scale up to system capacity and the *ISI* scheme does not perform well in a system with a large number of scope distributions, the *GBVC* method shows the best stability and scalability. Moreover, we observe that when the data update rate is much higher than the client hand-off rate, location-dependent cache invalidation may not be necessary.

For future work, various invalidation schemes can be evaluated under a one-dimension cellular model and a non-uniform user movement pattern etc. Moreover, a location-dependent cache invalidation scheme which considers temporal-dependent update frequencies is currently under investigation.

Acknowledgment

The authors would like to thank Dr. Tin-Fook NGAI for his valuable comments.

References

1. A. Acharya, B. R. Badrinath, T. Imielinski, and J. C. Navas. A www-based location-dependent information service for mobile clients. In *The 4th International WWW Conference*, January 1994.
2. B. R. Badrinath and T. Imielinski. Replication and mobility. In *Proc. of the 2nd Workshop on the Management of Replicated Data*, pages 9–12, 1992.

3. D. Barbara and T. Imielinksi. Sleepers and workaholics: Caching strategies for mobile environments. In *Proceedings of SIGMOD'94*, pages 1–12, May 1994.
4. B. Y. Chan, A. Si, and H. V. Leong. Cache management for mobile databases: Design and evaluation. In *Proceedings of 14th ICDE*, pages 54–63, 1998.
5. M. H. Dunham and V. Kumar. Location dependent data and its management in mobile databases. In *Proceedings of 9th International Workshop on Database and Expert System Applications*, pages 414–419, 1998.
6. Q. L. Hu and D. L. Lee. Cache algorithms based on adaptive invalidation reports for mobile environments. *Cluster Computing*, 1(1):39–48, Feb. 1998.
7. Q. L. Hu, D. L. Lee, and W.-C. Lee. Data delivery techniques in asymmetric communication environments. In *Proceedings of MobiCom'99*, August 1999.
8. J. Jing, O. Bukhres, A. K. Elmargarmid, and R. Alonso. Bit-sequences: A new cache invalidation method in mobile environments. *MONET*, 2(II), 1997.
9. V. Persone, V. Grassi, and A. Morlupi. Modeling and evaluation of prefetching policies for context-aware information services. In *Proceedings of MobiCom'98*, pages 223–231, October 1998.
10. H. Schwetman. *CSIM user's guide (version 18)*. MCC Corporation, http://www.mesquite.com, 1992.
11. S Shekhar, A. Fetterer, and D.-R. Liu. Genesis: An approach to data dissemination in advanced traveler information systems. *IEEE Data Engineering Bulletin*, 19(3):40–47, 1996.
12. K.-L. Wu, P. S. Yu, and M.-S. Chen. Energy-efficient caching for wireless mobile computing. In *Proceedings of 12th ICDE*, pages 336–345, Feb. 26-March 1 1996.

An Autonomous Data Coherency Protocol for Mobile Devices

Masahiro Kuroda[1], Takashi Sakakura[1], Takashi Watanabe[2], Tadanori Mizuno[2]

[1] Information Technology R&D Center, Mitsubishi Electric Corporation, 5-1-1 Ofuna, Kamakura, Kanagawa, 247-8501 Japan
{marsh,sakakura}@isl.melco.co.jp
[2] Shizuoka University, 3-5-1 Johoku, Hamamatsu, Shizuoka, 432-8011 Japan
{mizuno,watanabe}@cs.inf.shizuoka.ac.jp

Abstract. Wireless networks are becoming popular in data communication and suitable for data sharing among users using multicast capability, but have hidden node problems. The optimistic consistency scheme has been established with respect to data consistency and availability in disconnected environment. This paper proposes a data coherency protocol for mobile devices. This protocol lessens data traffic in data collaboration using multicast communication feature and is capable of automatic recovery from unpredictable disconnection, such as hidden node problem, by data versioning using version vectors and update log management in an optimistic data consistency scheme. We evaluated efficiency of data sharing over multicast communication and the recovery cost from failure caused by hidden nodes, and confirmed the efficiency of the protocol.

1 Introduction

The advance of the mobile computing infrastructure enables "anytime, anywhere" collaborative information services. These services need to ensure data coherency among mobile devices sharing information in wireless networks. Currently various wireless networks are available for mobile computing, but they are less reliable, have higher latency, and more costly than wired networks. There are two main approaches to solve these issues. One approach is to design a light-weight wireless transport protocol and translate the protocol into a wired protocol using a proxy model [1]. Second approach is to cache data in a site and send the differential data between two sites to the destination site. As for the latter case, there is an optimistic data consistency scheme [2, 3], with respect to data consistency and data availability. The traditional optimistic consistency scheme, such as those deployed in Coda [2] and Ficus [3], were basically aimed to offer data consistency and availability for devices using reliable and low latency networks, such as LAN. These schemes were also targeted to Unix file systems.

The timestamped anti-entropy protocol [5] uses data versioning and can be applied to various types of data. The protocol is based on replica maintenance using a version vector, an acknowledgement vector, and message logs. Each node exchanges the version vector periodically with its peer node and keeps the latest

H.V. Leong et al. (Eds.), MDA'99, LNCS 1748, pp. 194–205, 1999.
© Springer-Verlag Berlin Heidelberg 1999

data in it. This protocol assumes nodes involved in the exchange have timestamps that are comparable among the nodes by having frequent data exchange and can be extended to nodes that do not have comparable timestamps. Bayou's anti-entropy protocol for update propagation [6] is close to [5], but has a support for conflict detection and resolution based on per-write dependency for database operations.

Our optimistic data consistency model [9] is close to the timestamped anti-entropy protocol. The model deploys data versioning and the interface of conflict reconciliation that is defined for various types of applications. The model is also independent of communication media for data replication. Multicast networks are originally suitable for data sharing among mobile users. An invalidation protocol for data caching applicable to cellular networks by a periodic broadcast message is presented in [17].

We propose a data coherency protocol based on our data consistency model with data snooping mechanism of wireless multicast communications to attain information sharing among mobile users. This protocol offers consistent data view to mobile users by solving hidden node problems and by involving a less network traffic. Mobile users can join an information-sharing group or leave the group without caring about connected or disconnected and recover data missed during out of communication.

The first section describes our data consistency model and data versioning as a base function for data snooping and data recovery of the Autonomous Data Coherency Protocol. The next section explains the data coherency protocol. The data snooping mechanism, operation modes, data recovery and the state transition of the protocol are discussed.

The rest of this paper describes a prototype implementation and its evaluation. In the evaluation, the data sharing performance is measured and discussed. We conclude with a summary of this paper and future perspective of the protocol.

2   Data Synchronization Mechanism

We designed the data coherency protocol on a data consistency model that is intended to support unstable wireless communications. Our model uses the suggestion of the Mobile Network Computer Reference Specification (MNCRS) [11]. SyncStore is a container of data objects, and data objects in the container are kept consistent between a pair of containers by exchanging version vectors. Each SyncStore has a version vector to maintain consistency states of the data objects and the SyncStore itself. SyncStore manages update logs and its differential data as well. Every modification to an object in a SyncStore from an application results in generating an update log and its differential data. A synchronization executing entity, the Synchronizer exchanges update logs between a pair of Sync-Stores using the version vectors.

2.1   Data Versioning

The data versioning using version vectors to detect concurrent update conflicts was proposed by Parker, etc. [12]. The data versioning deployed in our model is similar to the timestamped anti-entropy protocol [5], but is based on an incremental counter. A version vector is a list, which is stored in a SyncStore, and is used for detection of concurrent update conflicts and for reduction of update logs.

A version vector contains elements of the form $(\text{ssid}_i, \text{lvcount}_i, \text{vcount}_i)$, where ssid_i is the identifier of the i-th SyncStore, lvcount_i is a version counter that is the incremental counter of its SyncStore, and vcount_i is a version counter that is the incremental counter of i-th SyncStore. To allow asynchronous data modifications on any node, which is not allowed in the protocol study [19], we extend the element by adding lvcount_i. If the local SyncStore is the i-th element, lvcount_i^i is incremented when there is an update in its local SyncStore and vcount_i^i is updated to the value of lvcount_i^i when a data synchronization is completed with another SyncStore. Therefore, the version vector is expressed as follows,

$$vv^i = vv_k^i \quad \{k = 1, 2, \ldots, n\}$$
$$vv_k^i = (\text{ssid}_k^i, \text{lvcount}_k^i, \text{vcount}_k^i) \tag{1}$$

The j-th component in the version vector of the i-th SyncStore means that the i-th SyncStore at the point of lvcount_j^i knows the number of the updates that were performed at the j-th SyncStore up to the point of vcount_j^i. The size n of the version vector vv^i infers that the i-th SyncStore knows n SyncStores including itself.

Assume each call to SyncStore's *markAsUpdated* method will cause its version vector to be appropriately updated and the update passed to the method will be logged. Each update logs has a field *ssid* and a field *lvcount* attached to it, where *ssid* indicates the SyncStore at which it was performed, and *lvcount* specifies the virtual time it was performed using the incremental counter.

Using the version vectors and update logs presented above, we can compute the difference of a SyncStores from another, detect concurrent update conflicts, and reduce the amount of update logs.

Let I_{ss} denote the i-th SyncStore that is initiating data synchronization, and R_{ss} denote the j-th SyncStore that is responding to the synchronization.

2.2   Difference Comparison

Assuming R_{ss} has received vv^i from I_{ss}, it then will compare with its vv^j.

If $\text{lvcount}_i^i > \text{vcount}_i^j$ and $\text{vcount}_j^j = \text{lvcount}_j^j$, I_{ss} is newer than R_{ss}. R_{ss} receives the differential data extracted from the update logs in I_{ss} after sending its vv^j as an acknowledgment.

If $\text{vcount}_j^i < \text{lvcount}_j^j$ and $\text{lvcount}_i^i = \text{vcount}_i^j$, R_{ss} is newer than I_{ss}. R_{ss} sets up differential data from its update logs by checking the *lvcount* that is attached to each update and is bigger than the lvcount_i^j, then sends the differential data to I_{ss}.

If $\text{lvcount}_i^i > \text{vcount}_i^j$ and $\text{vcount}_j^i < \text{lvcount}_j^j$, it is assumed that a concurrent update occurred. R_{ss} reconciles the conflict by invoking its application and brings the reconciled result into next data update.

If $\text{vount}_j^i = \text{lvcount}_j^j$ and $\text{lvcount}_i^i = \text{vcount}_i^j$, updates occurred neither on I_{ss} and R_{ss}, the data synchronization does nothing.

After each cycle of the synchronization, I_{ss} will compute the minimum *lvcount*, $v_{\text{lvmin}}{}^i$.

$$v_{\text{lvmin}}{}^i = \min\{\text{lvcount}_k^i | k = 1, \ldots, n\} \tag{2}$$

The i-th SyncStore can reduce its update logs by discarding the update logs with older *lvcount* than $v_{\text{lvmin}}{}^i$.

3   Autonomous Data Coherency Protocol

We define the Autonomous Data Coherency Protocol (ADCP) assuming a multicast data communication medium, such as wireless LAN as a shared data bus of centralized shared- memory multiprocessor computer with simple cache coherency mechanism based on snooping [16]. Nodes monitor data transactions on the bus and get necessary data from the bus. We call this function as *data snooper*. The *data snooper* takes data synchronization messages autonomously giving no side-effect on data synchronization sessions, then the synchronizer applies the messages into its SyncStore if the messages match requirements of the node. From the autonomous synchronization behavior, we call this operation mode of the ADCP as *sneak mode*.

Once all the nodes have the same data, no updates log is required as long as updates executed by nodes in the mode are serialized in a manner of accessing a shared bus. They receive update data, which is serialized by a medium, on a multicast data communication bus. We call this operation mode as *declared mode*.

3.1   Data Snooper

The *data snooper* is configured as a part of medium access control. The synchronizer uses a datagram interface as its transport communication to allow the *data snooper* monitoring the data synchronization messages. It prepares a port to listen for a data synchronization request from others, and a port for messages from the *data snooper*.

A list of data sharing nodes which is deducted from the version vector is configured into the *data snooper* via an interface of the datagram transport by the synchronizer. All data on the multicast communication medium is monitored if the sharing nodes send data synchronization messages on that medium. The *data snooper* queues the data to the monitoring port as a data synchronization message if the data is a data synchronization message.

We implemented the *data snooper* on Ethernet by enhancing the device driver to check data in the interrupt routine, and successfully tested that the data snooper could get a message from an IEEE 802.11 station via an Ethernet-802.11 bridge.

3.2 Sneak Mode

The synchronizer executes the following processes in the *sneak mode*. Let S_{ss} denote the k-th SyncStore that gets a synchronization message by snooping data exchanged between I_{ss} and R_{ss}. Let nvrs_i^j denote the number of increments supplied by I_{ss} to R_{ss}, and nvrs_i^k denote the number of logs applied to S_{ss}. S_{ss} calculates nvrs_i^j and nvrs_i^k as follows.

$$\mathrm{nvrs}_i^j = \mathrm{lvcount}_i^i - \mathrm{vcount}_i^j$$
$$\mathrm{nvrs}_i^k = \mathrm{lvcount}_i^i - \mathrm{vcount}_i^k \tag{3}$$

S_{ss} processes subsequent update log messages as follows.

1. $\mathrm{nvrs}_i^k{=}0$
   Drop the messages.
2. $\mathrm{nvrs}_i^j > \mathrm{nvrs}_i^k > 0$
   Drop the old ($\mathrm{nvrs}_i^j - \mathrm{nvrs}_i^k$) update log messages, and register the rest of messages. The sequence of the messages is checked by the *lvcount* in update logs. The *vcount* and *lvcount* of I_{ss}, R_{ss} and S_{ss} are updated by the applied S_{ss}' *lvcount* in its version vector to the new version.
3. $\mathrm{nvrs}_i^j = \mathrm{nvrs}_i^k$
   Register the messages, and update the *vcount* and *lvcount* of I_{ss}, R_{ss} and S_{ss} with the applied S_{ss}' *lvcount* in its version vector to the new version.
4. $\mathrm{nvrs}_i^j < \mathrm{nvrs}_i^k$
   Register the messages as "future" update logs without applying to the data. A lacked *vcount* of log will be retrieved from another node later by requesting with the missed *vcount*. When S_{ss} acquires a complete series of update logs and the lacked *vcounts* of update log, all logs are applied to the data.

If either $\mathrm{vcount}_k^i < \mathrm{lvcount}_k^k$ or $\mathrm{vcount}_k^j < \mathrm{lvcount}_k^k$ is true, assume a conflict occurred either between I_{ss} and S_{ss} or between R_{ss} and S_{ss}, the conflict is resolved by application program. If both updates are applied to S_{ss}, one update is applied at this synchronization and the other reconciled update is applied next time, when S_{ss} synchronizes with another SyncStore.

The new version of a snooped node will be reflected to the other nodes by executing data synchronization with them.

3.3 Declared Mode

When data update occurs at a node in declared mode, data message is sent to the shared nodes without generating update logs. The nodes in the session take the message and apply to its copy of data with no condition. An explicit data sharing session is established by a declaration message and the transition protocol from *sneak mode* to *declared mode*.

Start of a shared session. A shared session starts with a declaration message that a chair node, the session starter sends. A node on that medium takes the message by the *data snooper*, and the synchronizer is notified of the message. The declaration message contains the *lvcount* $V_{lvmin}{}^i$. $V_{lvmin}{}^i$ is the oldest *lvcount* in the version vector in the chair node. On either sending or detecting a declaration message, a node gets into another state ceasing both data synchronization and data snooping. Nodes detected the message try to respond to the node with the first lacked *vcount* or the expected *vcount* of update log in its i-th entry of version vector. The chair node waits for timeout from the last responded message. Then the chair node determines v_{min} that is the initial differential data supplied to responding nodes, from the collected *vcounts* of the chair node as follows;

$$v_{min} = \min\{\text{vcount}_i^r | r = 1, i - 1, i + 1, \ldots t\}$$
where r is id in number t of responded nodes.   (4)

After having determined v_{min}, the chair node sends the differential data from the next of v_{min} to the *lvcount* applied to its SyncStore.

All responded nodes send an acknowledgement message after receiving the data. Those nodes drop the differential data if it is already stored in their Sync-Store. All nodes move to the *declared mode* after the completion of this process.

As an optimized function, the session establishing process can reduce the update logs in respective SyncStores. All nodes can assume that all nodes are synchronized at the establishment of the session. Therefore all the nodes except the chair node can discard the log. The chair node still keeps the log for nodes, which missed the session.

Updating Data. Under the *declared mode*, data is updated by sending an acquiring message to the chair node. A "data updating" state is registered in all nodes of the session by acquiring the access right, therefore no other nodes can update the shared data. Once a node has succeeded in acquiring the access right, the synchronizer allows its application to update the data by releasing a lock for the SyncStore. The data management mechanism generates the differential data but not to generate update logs, and directly reflect the generated differential data to the data.

The synchronizer receives differential data in a process of executing the update and sends a data synchronization message with sequential id in the *declared mode* session instead of the *lvcount*. The other nodes receive the synchronization message and apply to their data without generating the update logs as well.

Cancellation of a Shared Session. The *declared mode* is cancelled by the chair node, which declared the start of the session. When the other nodes receive a cancellation message, make one version of update log since the declared session has started, and go to the *sneak mode*.

If a node in the *declared mode* notices that some version of differential data is missing by checking the sequential ids of updates, it moves to a disabled state and waits for the cancellation of the session. A node that has some problems in the session establishing process also goes to the state.

3.4 Missing Differential Data and Its Recovery

If a node missed some differential data, it sends a request message for the oldest missing data. A responder, who is the winner of acquiring the medium, responds to the request by the data with its version. The version of differential data provided by the responder is taken by not only the requested node but also nodes that missed the data too. When there is a conflict to apply the received differential data, a reconciliation procedure in application is executed on the data. The reconciliation procedure is repeated until the newest missing version of differential data. In some case, depending on an application to reconcile conflicts, a series of reconciliation may make another version of differential data. The data consistency is recovered by synchronization process eventually.

4 Evaluation

4.1 Evaluation System and Application

We deployed Ethernet as a communication medium with multicast capability. 16 client nodes are connected to the Ethernet assuming that a small number of data sharing nodes is involved in collaboration. A gateway server is also connected to the network, and a database server is connected to the clients via the gateway. The ADCP software is written in Java to have a uniform infrastructure for all the devices. ADCP software is written in Java and installed on all the clients and the gateway.

We use a voting system for nomadic auctions as an application and assume 16 attendants including a broker in the system. Tables of product records represent auction items and are stored in the database server. Each record has 16 fields to register auction prices by attendants and an auction status field whether to set "open" or "close" by a broker. The broker retrieves a table from the database server via the gateway per auction session, creates a data object representing the table and registers it into its SyncStore. The object equips methods for accessing data at every SQL command level such as "update" or "select". Differential data of the object is extracted as a SQL statement as well.

Each session attendants set ups a SyncStore where data sharing relations to all other nodes are profiled. The broker declares the beginning of the session, once the initial data object is delivered to each SyncStore via the declaration establishing process.

Any attendant can price a product at any time. A pricing operation is reflected to the other attendants by extracting the differential data as an SQL statement and multicasting it in the ADCP. Once an attendant gets out of the session, no more operation is allowed. He or she can join the next session and also knows the result of the previous session.

The broker closes the auction session after setting the "close" state to all records in the object. The broker sends the cancellation message of the *declared mode*, and writes back the differential data of the object, that is the result of the session, to the backend database server via the gateway.

Table 1. Response time comparison

Synchronization mode	Response time
Pair-wise	650.7 millisecond
Sneak	523.0 millisecond
Declared	26.7 millisecond

4.2   Methodology of Evaluation

We measured the response time for reflecting a data modification to all nodes in each synchronization mode. For the recovery cost evaluation, it might be obvious that the ADCP can lessen communication traffic comparing to a pair-wise data synchronization if the number of data sharing nodes is more than two. However, problems in mobile environment lead to more communication traffic in running the ADCP for the recovery. Therefore, we measured the amount of communication traffic in the application while making a hidden node situation by disconnecting clients from the Ethernet. Moreover, we compared the data recovery costs in the *declared mode* with that of the *sneak mode*.

4.3   Measurement Conditions

All updates result in generating 100 bytes-length SQL statements in differential data, and subsequent synchronization delivers the data to others and recovers the consistency.

We iterated update-and-synchronization 100 times and varied the number of attendants from 2 to 16 to see how the node number affects the communication traffic in the *sneak* and *declared mode*, while one of the attendants is always the broker.

The cost for delivering the initial data to attendants is included in the result. The initial data (1924 bytes Java object) contains 10 records of the products. Each modification generates 100 bytes-length differential data.

The auction session lasts t seconds. Data modifications on nodes follow the Poisson distribution, predetermined number of nodes being hidden at the time $t/2$, and data recovery is executed after closing the session in this hidden node simulation.

Additionally, attendants execute data synchronization after each data modification, and the broker executes data synchronization with a proper node to deliver status information change in the *sneak* mode, and the recovery process is always held in the *sneak mode*.

5   Result of Measurements

5.1   Response Time

Table 1 shows the response time in each mode. Total elapsed time for reflecting an update is measured while fixing the number of nodes to three to simplify the comparing condition. Two synchronization operations are used in the *pair-wise mode* and one synchronization operation is needed in the *sneak mode*. The time in the *declared mode* is more than 20 times faster than the other modes. On the other hand, the response time in the *sneak mode* is still close to the *pair-wise mode* despite of the fact that the number of synchronization per update is half of the pair-wise mode. These results suggest that the cost for logging is high comparing to the version vector exchange.

5.2   Communication Traffic

Figure 1 shows the comparison between the declared mode and the sneak mode, while the X-axis presents the number of attendants and the Y-axis is the total bytes of the traffic. The traffic in the *declared mode* stays low as expected, whereas the traffic in the *sneak mode* rises rapidly as the number of node increase.

To configure the ADCP working, each node needs to have the data sharing relations with other nodes managed in the version vector. Data modifications in the *sneak mode* result in sending the version vector aside from the differential data as the data synchronization process.

The cost for exchanging the version vector, $\text{Cost}(n)$, is estimated as; $\text{Cost}(n) = (64n+112)(n+1)+428$, where n presents number of nodes. The complexity of the cost is $O(n^2)$, and the number of nodes is the dominant factor of communication traffic in this evaluation.

This result suggests that the number of data sharing nodes in the *sneak mode* are limited by the cost for exchanging version vector comparing to the *declared mode*.

5.3   Communication Traffic with Hidden Nodes

Total data traffic among the broker and 15 attendants is measured as Figure 2. The number of hidden nodes that are disconnected from the network in the middle of the session is presented by the X-axis. The traffic includes the communication traffic by 15 attendants, and the traffic for data recovery by respective number of hidden nodes. Both modes show the same tendency in the total data traffic. The constant difference between the two modes is assumed as the version vector sending cost from the result of Fig. 1. The total traffic is proportional to the number of hidden nodes.

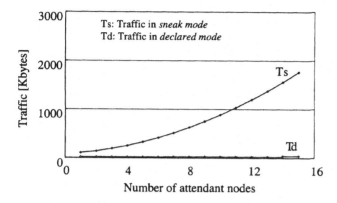

Fig. 1. Communication traffic

5.4   Estimated Recovery Cost

Data recovery is processed after the session close. In the case of the *declared mode*, the state of node changes from DS to SNP, then the recovery is processed in the *sneak mode* so that the message for data recovery is shared by failed nodes in the *sneak mode* manner. (Ts–Tsc) and Td are the traffic for data recovery, and no significant difference in tendency between two modes is observed. The recovery cost is described as; Recovery cost = Nh · Cost(16) + C_{miss}, where C_{miss} denotes the total amount of missed differential data, and is relatively small. N_h expresses the number of hidden nodes in the session. The complexity of recovery cost is $O(N_h)$, so that the recovery cost for hidden nodes in traffic increases only linearly if the number of hidden nodes increases.

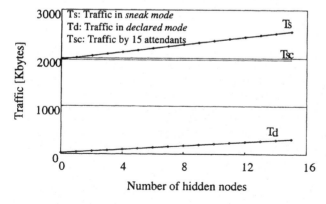

Fig. 2. Communication traffic with hidden nodes (Number of attendant nodes = 15)

6   Conclusion

We introduced a data coherency protocol ADCP that deploys a data synchronization mechanism and a data snooping mechanism on a multicast data network. This protocol allows mobile devices to disconnect unpredictably and keep data shared consistent among the devices. We evaluated the practicability of the protocol by estimating synchronization response time and data recovery cost in data traffic from failures caused by such as hidden node problem. The methodology used in the estimation was simple, however we confirmed that the protocol reduced data traffic and brought better performance in response.

We will further evaluate the protocol in aspects of medium acquiring cost in a large number of nodes and recovery cost in practical hidden node situations. We will also augment the protocol to meet large-scale system requirements and multi- hop communication environment.

Acknowledgements

We would like to thank Mr. Ryoji Ono and Mr. Tatsuji Munaka for their indispensable contributions to this work, especially in the evaluation efforts and data preparation.

References

1. B. Zenel and D. Duchamp, General Purpose Proxies: Solved and Unsolved Problems. In Proceedings of Hot-OS VI, May 1997
2. Mahadev Satyanarayanan, et al, "Coda: A Highly Available File System for a Distributed Workstation Environment", IEEE Transactions on Computers, 39(4), April 1990.
3. Richard G. Guy, et al, "Implementation of the Ficus Replicated File System", USENIX Conference Proceedings, USENIX, June 1990.
4. C.H. Papadimitrio, "The Serializability of Concurrent Database Updates", Journal of the ACM, 26(4), October 1979.
5. Rechard A.Golding, "Weak consistency group communication and membership," Ph.D. thesis, published as technical report UCSC-CRL-92-52. Computer and Information Sciences Board, University of California, Santa Cruz, December 1992.
6. Douglas B. Terry, et al,"Managing Update Conflicts in Bayou, A Weakly Connected Replicated Storage System", In Proceedings of the 15th ACM Symposium on Operating Systems Principles, ACM, 1995.
7. Sunil K. Sarin, Nancy A. Lynch, "Discarding Obsolete Information in a Replicated Database System,"IEEE Transactions on Software Engineering, SE-13 (1), pp.39–47, Jan. 1987.
8. Karin Pertersen, Mike J. Spreitzer, et al, "Flexible Update Propagation for Weakly Consistent Replication", Proceedings of the 16th ACM Symposium on Operating Systems Principles (SOSP-16), Oct. 1997.
9. Masahiro Kuroda, et al,"Data Transfer Evaluation of Nomadic Data Consistency Model for Large-scale Mobile Systems",IEICE Transactions on Information and Systems, Vol.E82-D, No.4 Apr. 1999.

10. Masahiro Kuroda, et al, "Wide-Area Messaging System on Nomadic Data Consistency", The 13th International Conference on Information Networking, January 1999.
11. Mobile network computer reference specification, http://www.mncrs.org
12. D. Stott Parker, et al, "Detection of Mutual Inconsistency in Distributed Systems", IEEE transactions on Software Engineering, 9(3), May 1983.
13. David Ratner, et al,"Dynamic Version Vector Maintenance", Technical Report CSD-970022, University of California, Los Angeles, June 1997
14. D. L. Schilling, S. G. Glisic and P. A. Leppanen(Ed); "Overwiew of wideband CDMA", Wireless Communications: TDMA vs. CDMA, Kluwer, 1997.
15. Jaap Haartsen, Mahmound Naghshineh, et al,"Bluetooth: Vision, Goals, and Architecture",ACM Mobile Computing and Communications Review, Volume 2, Number 4.
16. D.A.Patterson, J.L.Hennessy, "Computer Architecture a Quantitative Approach Second Edition", Morgan Kaufmann Publishers, Inc..
17. Daniel Barbara, Tomasz Imielinski, "Sleepers and Workaholics: Caching Strategies in Mobile Environments" , ACM SIMOD 94- 5/94 Minneapolis
18. Yixiu Huang, et al, "Data Replication for Mobile Computers", ACM SIGMOD 94- 5/94 Minneapolis
19. Masahiro Kuroda, et al, "A Study of Autonomous Data Coherency Protocol for Mobile Devices", to be appeared in proceedings of the ICCCN'99, October 1999

Session V: Mobility and Location Management

Location Management Strategies for Reducing Both Movement Cost and Locating Cost Simultaneously

Chao-Chun Chen and Chiang Lee

Department of Computer Science and Information Engineering
National Cheng-Kung University
Tainan, Taiwan
leec@dblab.iie.ncku.edu.tw

Abstract. A major problem in a wireless communication system is how to locate mobile clients. This is named the *location management* issues. Two major operations are involved in managing a mobile client's location: the movement operation and the locating operation. The past methods can only minimize the cost of one of these two operations, but not both. The major contribution of this paper is to propose strategies that minimize the costs of both operations simultaneously. Our performance analysis proves that the proposed strategies are superior to the past methods.

1 Introduction

A great difference between wireless communication environment and traditional wireline communication environment is that it allows users migrating anywhere and retrieving all kinds of data anytime in mobile computing environment. Figure 1 shows the architecture of a wireless communication system. To effectively monitor the movement of each mobile client, a large geographical region is partitioned into small registration areas (RA). Each RA has a mobile switch center (MSC) which serves as the local processing center of the RA. The profiles of mobile clients inside a RA are kept in the MSC's visitor location register (VLR). On top of several MSC/VLRs is a local signaling transfer point (LSTP) and on top of several LSTPs again is a remote signaling transfer point (RSTP). The LSTP and the RSTP are routers for handling message transfer between stations. For one RSTP, there is a home location register (HLR). Each mobile client must register in a HLR. In this way, the whole system forms a hierarchy of stations.

The major problem in a wireless communication system is how to locate a mobile user in a wireless environment. The *Basic HLR/VLR strategy* is most often referred in resolving this problem. IS-41[IS-91] used in the United States and GSM[MP92] used in Europe are examples of this strategy. Many papers in the literature have demonstrated that the Basic HLR/VLR strategy does not perform well. This is mainly because whenever a mobile client moves, the VLR

H.V. Leong et al. (Eds.), MDA'99, LNCS 1748, pp. 209–219, 1999.
© Springer-Verlag Berlin Heidelberg 1999

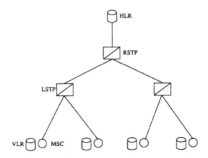

Fig. 1. Architecture of a wireless communication system.

of a RA which detected the arrival of the client always reports to the HLR about the client's new location. For convenience in presentation, we will simply use VLR as a general term which represents the local hardware/software system managing mobile clients information within a RA. While a call is placed, the callee is also located by going to the HLR's database to find the callee's new location. As the HLR could be far away from a VLR, communication to the HLR is costly. Several other location management strategies have been proposed to improve the performance of the Basic strategy. Among them, the Forwarding strategy [JLLM95,CCL98,CCL99], the Local Anchor (LA) strategy [HA96], and the Caching strategy [JLLM94] are representatives of the proposed methods.

Our strategies, termed the Forwarding on Local Anchor (FLA) strategies, try to further improve the performance of these past methods by minimizing both the movement cost and the locating cost at the same time. The key idea is to allow multiple LAs that a mobile client has traveled through to be linked together by using forwarding links. If a call is from any VLR overseen by one of these linked LAs, the call can be quickly forwarded to the callee's currently residing VLR through the forwarding links of LAs. The callee's location will be recorded and can be located in this way as long as the call is from a VLR that the callee has ever visited. In this way, locating cost is reduced because the HLR database need not be accessed as long as the callee can be located through the LA forwarding links. Movement cost is also reduced by using this method because update of a client's location need not be reported to the HLR as long as the mobile client has not moved out of the region of a LA (i.e., the regions of all the VLRs overseen by this LA).

To differentiate all the above mentioned methods, we classify them based on two dimensions, as shown in Figure 2. One dimension is the HLR update frequency of movement operations and the other is the HLR access frequency of locating operations. They stand for respectively the needed updates and accesses to the HLR database while a mobile client moves to a new location and is called by a caller. Our proposed FLA strategies combine the features of the other methods and minimize both HLR access and HLR update frequencies in the operations.

In this paper, a Static FLA strategy and a Dynamic FLA strategy are proposed. The algorithms will be presented in detail and relevant issues will also

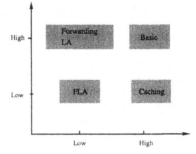

Fig. 2. The classification of various location management strategies.

be discussed. In order to compare the performance of these strategies, we derive the cost models of the past methods and a thorough analysis is conducted. In the evaluation, we find that a dramatic performance improvement over the past methods is achieved by using the proposed FLA strategies. Note that because of size limitation, some subtle issues of the algorithms, cost models, and performance results cannot be presented. The reader may refer to [LC99] for details.

The remainder of this paper is organized as follows. In Section 2, we presents the proposed FLA strategy. Then the cost model of each strategy is derived in Section 3. The evaluation results are given in Section 4. Finally, we summarize the paper and describe our future work in Section 5.

2   The FLA (Forwarding on Local Anchor) Strategy

2.1   The Supporting Data Structure

TO put our proposed strategy into practice, a data structure named the Mobile User Landmark (MULM) table is defined to save the visited mobile users of a VLR. As shown in Figure 3, this table maintains for each visited mobile user the user's ID (MU_{id}), a *Type*, and a *Pointer*. The MU_{id} has the identifier of the user.

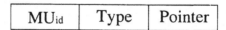

Fig. 3. The schema of the MULM table in VLRs.

The second attribute *Type* is to tell the type of the mobile user's belonging VLR while the mobile user is in this VLR. There are four types for each VLR. They are used together with the third attribute, a pointer, to locate the callee. The four types are the *Latest VLR*, the *Latest LA*, the *Visited VLR*, and the *Visited LA*. We describe the Type and the Pointer values as well as their meanings in the following.

1. Latest VLR: A *Type* value of "Latest VLR" means that this mobile user is right at the RA managed by this VLR (or simply *at this VLR* for convenience). We use a NULL to represent that it is the end of a chain of VLRs.
2. Latest LA: The value "Latest LA" represents that this mobile user is currently at the VLR to which the *Pointer* is pointing.
3. Visited VLR: The value "Visited VLR" means that this mobile user had visited this VLR and is not here at this moment. The *Pointer* value records the LA while the mobile user was at this visited VLR.
4. Visited LA: The value "Visited LA" means that this mobile user had visited this VLR and has left. And, while the mobile user is at the VLR, this VLR is also the LA. The *Pointer* value records the next LA that the mobile user moves to.

2.2   The Static Fla Strategy: Movement Operation

We formally present the algorithms in the following subsections. First, we present the movement operation of the static FLA strategy. Basically, tasks of a movement operation are to save a record in the VLR describing the new VLR that the mobile user has gone to and to update the old VLR in the record of the corresponding LA to the new VLR.

The movement operation of the static FLA strategy:

Figure 4 gives the tasks when a mobile user moves to a new RA. Same as all the other figures in this section, the solid lines always represent messages transmitting between the RAs, and the broken lines are to represent the location to which the *Pointer*'s value refers.

1. The new VLR learns that the mobile client is inside its territory and informs the old VLR that the mobile client is in its RA. The MULM table of the new VLR is inserted a new record, (MU_{id}, "Latest VLR", NULL), describing the coming mobile client. If the mobile client visited this new VLR in the past, then the system only updates the *Type* and the *Pointer* values.
2. The old VLR replys an acknowledgement to the new VLR.
3. The old VLR informs the LA that the mobile client has moved to the new VLR. Also, the old VLR updates its own MULM table by replacing the mobile client's *Type* value with "Visited VLR" and the *Pointer* value with the LA's location.
4. The LA replys a message to the old VLR, and updates its own MULM table. The *Type* value is not changed, and the *Pointer* value is changed to the new VLR of the mobile client.
5. End of the movement opreation.

2.3   The Dynamic Fla Strategy: Movement Operation

The basic idea of the dynamic FLA strategy is that the new VLR determines whether reports to the LA or the HLR the mobile client's new location when a mobile client comes into its territory. Figure 4 gives the process that the new VLR reports to the LA, and Figure 5 tells how the new VLR reports to the HLR about the mobile client's new location. The algorithm is as follows.

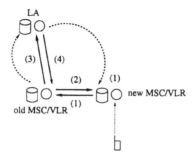

Fig. 4. Scenario 1: The new VLR informs the LA of the mobile client's new location.

The movement operation of the dynamic FLA strategy:

1. When the new VLR learns that a mobile client is coming into its territory, it will determine whether requesting a location update to the LA or the HLR. If the new location of the mobile client is outside the area that the RSTP of the original LA controls before the mobile client moves, jump to the Step 7. Otherwise, go on to the next step.
2. Step 2 ~ Step 5 are the same as those in the static FLA strategy presented above.
6. Goto Step 13.
7. The new VLR informs the HLR of the arrival of the mobile client.
8. The HLR returns an acknowledgement to the new VLR, and the new VLR's MULM table is inserted a record, (MU$_{id}$, "Latest LA", NULL).
9. The HLR updates its data about the mobile client's new location and informs the old LA about the movement of the mobile client.
10. The second and third attribute values in the MULM table of the old LA become "Visited LA" and the new VLR's location (which is the new LA). The old LA also sends an acknowledgement message to the HLR.
11. The old LA informs the old VLR of location change of the mobile client.
12. The second and third attribute values in the old VLR's MULM table are changed to "Visited VLR" and the location of the old LA, respectively. The old VLR also sends an acknowledgement to the old LA.
13. End of the movement opreation.

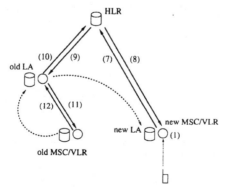

Fig. 5. Scenario 2: The new VLR informs the HLR of the location update.

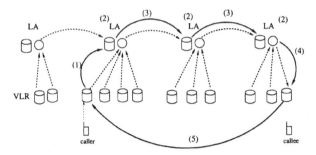

Fig. 6. Scenario 1: Locating the callee through linked LAs.

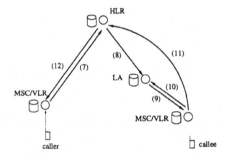

Fig. 7. Scenario 2: Locating the callee through HLR.

2.4   The Fla Strategy: Locating Operation

Figure 6 and Figure 7 are two cases that could occur while locating the callee. Figure 6 shows how to locate the callee through the chain of LAs that the callee have visited in the past. Figure 7 shows the other case that the system locate the callee through HLR. The locating operations in the static FLA and the dynamic FLA strategies are the same. The details of the operation is as follows.
Locating operation of the FLA strategy:

1. When a VLR receives a request of locating a callee, it first checks whether its MULM table has the callee's record. If yes, then sends the locating request to the LA stated in this record (on the *Pointer* field). Otherwise, jump to Step 7.
2. If the record of the MULM table of the LA stated in Step 1 says that this LA is a "Visited LA", then goto Step 3. If it says "Latest LA", then goto Step 4.
3. The locating request is forwarded to this visited LA. While the request is forwarded to the next LA, the callee's record is again searched from this LA's MULM table. Go to Step 2.
4. The latest LA finds the callee's record from the MULM table. If the value of the "Pointer" field is NULL, then the callee is right in one of this LA's governing RAs. Hence, a message is forwarded to the caller's VLR to make the connection. Goto Step 13. If the value of the "Pointer" field is not NULL, then the locating request is sent to the latest VLR to which the "Pointer" field refers.
5. The latest VLR sends a message to the caller's VLR to make a connection.
6. Goto Step 13.

7. The caller's VLR sends the locating request to the HLR.
8. The HLR forwards the request to the callee's LA.
9. The callee's latest LA forwards the request to the lastest VLR. Also, the callee's record in this VLR's MULM table is updated by replacing its MULM.Type with "Latest LA" and MULM.Pointer with NULL.
10. The callee's VLR acknowledges the receipt of the message to the LA and the LA will then update the callee's record in its MULM table by replacing MULM.Type with "Visited LA" and MULM.Pointer with a pointer to the callee's current residing VLR (i.e., the latest LA).
11. The callee's VLR sends a message to the HLR. The HLR updates the callee's new location to the new latest LA.
12. The HLR forwards the message about the current VLR of the callee to the caller's VLR and the connection between the caller's VLR and the callee's VLR is build.
13. End of the locating operation.

3   Cost Model

3.1   Parameters

The parameters used in our cost models are listed in Figure 8. The costs of the Basic HLR/VLR strategy, the Forwarding strategy, and the LA strategy were discussed in the literature[MP92,JLLM95,CCL98,HA96]. But the environments and the details that were referenced in their derivations are different in many ways. In order to make a fair and reasonable comparison, we make some general assumptions and based on which, we derive their cost functions in a uniform way.

Symbol	Meaning
h_1	The cost of sending a message from a VLR to another VLR, and these two VLRs are in different RSTPs.
h_2	The cost of sending a message from a VLR to another VLR through a RSTN;
h_3	The cost of sending a message from a VLR to another VLR through a LSTP;
P_L	The probability of a mobile client's moving into a new RA which is under the same LSTP as the last RA that the client just left;
P_R	The probability of a mobile client's moving into a new RA which is under the same RSTP as the last RA that the client just left;
CMR	The call-to-mobility ratio;
C_Y^X	The cost of performing X in strategy Y, where $X \in \{M, L, total\}$ and $Y \in \{Basic, SLA, DLA, Fwd, SFLA, DFLA\}$.
$C_{FLA-link}$	The communication cost between two VLRs through a forwarding link
k_{FLA}	The length of a forwarding link in using the FLA strategy
ρ_i	The probability of a caller's request is issued from LA_i and its overseeing VLRs

Fig. 8. The symbols of the parameters and their meanings.

3.2   Derivation of Cost Functions

As the tasks of location management include managing movement operations and managing locating operations, it is natural to calculate the costs from these two sets of operations. As the ratio of the number of calls (locating a callee) to mobility (movement of a callee) is represented by CMR, the total cost can be normalized and defined as

$$total\ cost = \frac{1}{CMR} \cdot movement\ cost + locating\ cost$$

Because of space limitation of the paper, we can only present the cost model derivation of the proposed Static FLA strategies in detail. For the cost models of the other strategies, the reader may refer to [LC99] for the detailed derivation.

Cost functions of the FLA strategy   The total cost of Static FLA strategy can be represented as

$$C_{SFLA}^{total} = (\frac{1}{CMR}) \cdot C_{SFLA}^{M} + C_{SFLA}^{L}$$

In the movement operation, we only count the costs of Step 1 $\sim$ Step 3 (referring to the algorithm presented in the last section), because Step 4 involves only an acknowledgement to a previous VLR which is not a critical task in the movement operation. In the algorithm, as Step 2 and Step 3 can obviously be executed simultaneously, the movement cost can be computed from Step 1 and Step 2 only. As for the movement of a mobile client from the old VLR to the new VLR, we find three possible situations:

- The new VLR and the old VLR are under the same LSTP: the expected movement cost of this situation is $P_L \cdot (2 \cdot h_3)$.
- The new VLR and the old VLR are under the same RSTP but different LSTPs: the expected movement cost is $P_R \cdot (2 \cdot h_2)$.
- The new VLR and the old VLR are under different RSTPs: the expected movement cost is $(1 - P_L - P_R) \cdot (2 \cdot h_1)$.

The addition of these costs gives the movement cost of the Static FLA strategy.

$$C_{SFLA}^{M} = P_L \cdot (2 \cdot h_3) + P_R \cdot (2 \cdot h_2) + (1 - P_L - P_R) \cdot (2 \cdot h_1)$$

For the locating operation, two cases are involved:

1. The callee is found through HLR.
2. The callee is found through a VLR where the callee visited before and is currently locateed.

The cost of the first case is the same as C_{SLA}^{L}. The probability of the occurrence of Case 1 is $1 - \sum_{i=0}^{k_{FLA}} \rho_i$, because the HLR is needed to locate a callee only when the callee is not in any of the VLRs that the client has visited. The cost of

Case 2 above is $C_{FLA-link}+i \cdot C_{FLA-link}+C_{FLA-link}+C_{FLA-link}$. By counting the cost of each case with its probability, we have the following total locating cost

$$C_{SFLA}^{L} = (1 - \sum_{i=0}^{k_{FLA}} \rho_i) \cdot C_{SLA}^{L} + \sum_{i=0}^{k_{FLA}} (\rho_i \cdot (C_{FLA-link} + i \cdot C_{FLA-link} + C_{FLA-link}$$
$$+C_{FLA-link}))$$

4 Performance Analysis

From the equations of the cost models in the last section, we see the important factors that affect the performance: the distance that a message is transferred. In the following, we evaulate the performance of different strategies based on the derived cost models. The default values of the parameters used in our evaluation are given in Figure 9. We show the effect of distance of message forwarding

Parameter	Default value
h_1	8
h_2	4
h_3	1
P_L	0.7
P_R	0.2
CMR	0.5
k_{FLA}	6
ρ_i	0.05

Fig. 9. Default values of parameters.

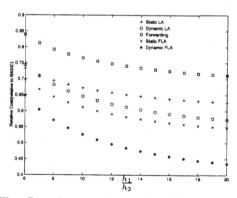

Fig. 10. The effect of message forwarding distance to performance.

to the cost of location management. The least communication cost is for those communication between two VLRs within the same LSTP (which is h_3). The highest communication cost is for those between a VLR and the HLR (which is h_1). Figure 10 shows the result under varying h_1/h_3. For every strategy, the cost ratio reduces when h_1/h_3 increases. This is because a large h_1/h_3 means that communicating with the HLR is very costly. All the strategies are designed to improve the Basic HLR/VLR strategy by reducing the effect of a high h_1 to the total cost. Hence, an increase of h_1/h_3 will indeed enlarge the performance difference between the other strategies and the Basic HLR/VLR strategy (that is, the curves drop). Also, we can see from this figure that the FLA strategy outperforms the others with a significant margin. The improvement over the Forwarding strategy is around 20%~30%, and is over the LA strategy by 15%~20%. The improvement is mainly due to the fact that the FLA strategy reduces both the movement cost and the locating cost simultaneously, while the Forwarding strategy and the LA strategy can only reduce one of the two costs. For both the LA and our FLA strategies, the dynamic version is slightly better than the static version as expected.

5   Conclusions

In this paper, we proposed a location management strategy, called the FLA strategy. Different from the past strategies, the FLA strategy can reduce both the movement cost and the locating cost at the same time. Two versions of the FLA strategies were intorduced, the static FLA strategy and the dynamic FLA strategy. We also derived the cost models of the proposed strategies and several other related works. Our analysis results reveal that in most case the static FLA strategy performs better than all the other strategies. Our future work is to study the issue that the path of a mobile client may contain cycles. In this situation, the forwarding links form cycles, which wastes traversal time while the system follows the links to locate a client. It can be complicated when a large number of mobile clients and their paths are considered.

References

[CCL97] Ing-Ray Chen, Tsong-Min Chen, and Chiang Lee. "Performance Characterization of Forwarding Strategies in Personal Communication Networks". In *21th IEEE International Conference of Computer Software and Application(COMPSAC'97)*, pp. 137-142, Washington, D.C., August 1997.

[CCL98] Ing-Ray Chen, Tsong-Min Chen, and Chiang Lee. "Performance Evaluation of Forwarding Strategies for Location Management in Mobile Networks". *The Computer Journal*, Vol. 41, No. 4, August 1998, pp. 243-253.

[CCL99] Ing-Ray Chen, Tsong-Min Chen, and Chiang Lee. "Analysis and Comparison of Location Strategies for Reducing Registration Cost in PCS Networks". *To appear in Wireless Personal Communications Journal*, 1999.

[HA96] Joseph S. M. Ho and Ian F. Akyildiz. "Local Anchor Scheme for Reducing Signaling Costs in Personal Communications Networks". *IEEE/ACM Transactions on Networking*, Vol. 4, No. 5, October 1996, pp. 709-725.

[IS-91] EIA/TIA IS-41.3. "Cellular Radio Telecommunications Intersystem Operations". Technical Report, Technical Report (RevisionB), July 1991.

[JLLM94] Ravi Jain, Yi-Bing Lin, Charles Lo, and Seshadri Mohan. "A Caching S-
trategy to Reduce Network Impacts of PCS". *IEEE Journal on Selected Areas in
Communications*. Vol. 12, No. 8, October 1994, pp. 1434-1444.

[JLLM95] Ravi Jain, Yi-Bing Lin, Charles Lo, and Seshadri Mohan. "A Forwarding
Strategy to Reduce Network Impacts of PCS". *IEEE INFOCOM* Vol. 2, No. 8,
1995, pp. 481-489.

[LC99] Chiang Lee and Chao-Chun Chen. "Location Management Strategies for Reducing Both Movement Cost and Locating Cost Simultaneously". Technical Report
#0629, 1999.

[MP92] M. Mouly and M.-B Pautet. "The GSM System for Mobile Communications".
In *Proceedings of 13th JSSST Conference*, Palaiseau, France, 1992.

Internet Mobility Support Optimized for Client Access and Its Scalable Authentication Framework

Sufatrio, Kwok-Yan Lam

Centre for Systems Security Research,
School of Computing, National University of Singapore,
Lower Kent Ridge Road, Singapore 119260
{sufatrio, lamky}@comp.nus.edu.sg

Abstract. Besides allowing a moving Mobile Node (MN) to keep receiving datagrams, the use of permanent home address in Mobile-IP also makes MN available to accept new connections from Corresponding Nodes. However, it's expected that in most cases MN only runs client sides of the applications that always initiate the connections. Given this observation and the often-disconnected nature of MN, the virtue of permanently associating MN with one fixed home address and Home Agent becomes questionable, while it already imposes potential inefficiency in its indirect routing. We argue that it's necessary to provide an option that lets client-access MN make use of multiple home addresses and Home Agents that are based on connection sessions. We propose a new enhancement scheme providing such a mechanism in this paper, which aims to let MN trade some additional lightweight local processing for better routing and seamless roaming. A further mechanism to ensure smooth handoff under this new scheme is proposed as well. Finally, we complete the paper by outlining a scalable authentication protocol for securing its operation, with the use of minimal public key cryptography.

1   Introduction

Supporting mobility within the Internet still remains as an active research area as it's indeed promising to realize the idea of having Internet always available even as we move. Much work has been done in this area, particularly under the effort of IETF with its Mobile-IP standard protocol scheme [1, 2]. In Mobile-IP scheme, a Mobile Node (MN) continues to be identified by a permanent home address and keeps receiving datagrams by help of a designated Home Agent (HA) on its home network.

With the use of this permanent home address and on stand-by HA, Mobile-IP also makes MN always available for any Corresponding Nodes (CNs) that might wish to initiate a new connection to it. However, it's expected that in most cases MN is used only to run client sides of applications to access services available on the Internet. It's unlikely that MN is deployed as some application server that

H.V. Leong et al. (Eds.), MDA'99, LNCS 1748, pp. 220–229, 1999.
© Springer-Verlag Berlin Heidelberg 1999

goes mobile, mainly due to the inherent possibility of disconnection imposed by current wireless networking technology and power supply issue.

From the assumption that MN only runs client sides of applications that always initiate the connections[1], it now can be assumed that all incoming datagrams to MN are sent in reply to previous MN's requests. In this new scenario, MN's home address and its corresponding HA are required to be retained the same only within each individual connection session. With the flexibility of not having to maintain home address and HA fixed at all times, we propose a new enhancement on Mobile-IP that's suitable for client-access type of MN. Under our proposal of "Session-Based Client-Access Mobile-IP", it's possible for MN to have multiple dynamically-allocated home addresses and HAs that might be different from one connection session to subsequent connection sessions.

This enhancement is indeed appealing to help facilitate Mobile-IP wide deployment due to its new advantages offered, namely better routing and seamless roaming. With the availability of this new scheme, we then envision that HA service may be offered as part of available resources on the foreign network. Therefore, HA can always be close enough to the current Foreign Agent (FA), preferably at the same sub-network. By following this principle of putting HA as close as possible to MN's present point of attachment, our enhancement provides "the achievable optimal path" to the problem of triangular routing. In addition, this scheme also generalizes the operation of dynamic home address allocation as specified in some latest Mobile-IP drafts [3, 4]. While actually performing the mechanism in [3, 4] already makes MN impossible to be known as destination end for CN-initiated connection, there is no mention in those documents on how MN may possibly gain further benefits.

While this scheme introduces some overhead on MN, the new lightweight processing should not be a computational burden for most MNs. However, to ensure smooth and fast handoff under this new scheme, a further mechanism that takes advantage of pre-generated registration messages is proposed as well.

Finally, as any protocol in mobile wireless environment needs a strong security protection, following the description of our new scheme, a security framework for securing its operation is then considered. We outline in this paper our scalable public-key based authentication protocol which sets only minimal computing requirement and administration cost on MN.

The remainder of this paper is organized as follows. Section 2 outlines the Session-Based Client-Access Mobile-IP. Further mechanism for ensuring fast handoff is the topic of Section 3, which is then followed by some discussions on performance aspect in Section 4. The scalable security framework is presented in Section 5. And finally, Section 6 closes the paper with conclusion.

---

[1] It is assumed that access to the application is provided by means of some "client-oriented" protocol, in which a client can initiate request sessions from any point of attachment. This assumption is in line with the current accepted practice and trend in developing Internet applications, such as POP for mail service access.

2   Session-Based Client-Access Mobile-IP

By allowing multiple home addresses to be associated with a MN, we generalize the current operation of Mobile-IP. While still maintaining interoperability, the enhancement improves routing efficiency of Mobile-IP by minimizing the overhead in its indirect routing. As a result, topologically correct reverse tunneling [5] is then acceptable for use to allow Mobile-IP coexist with ingress filtering.

This new scheme relies on mechanisms by which MN can learn new available HAs that are close to it. In the next following two subsections, we describe two ways in which this can be done. Section 2.1 outlines a method called Home Agent Relocation, while Section 2.2 describes the dynamic allocation of home address and HA on the visited foreign network. Following that, Section 2.3 specifies some new considerations at MN and FA.

2.1   Home Agent Relocation

Home Agent Relocation mechanism can be initiated by HA to inform its client-access MN of other available HAs that may better support it. This may be desirable, for example when HA learns that MN' s COA is located too far and it knows of some other HAs that can give better routing. A heavy-loaded HA may also employ this mechanism for load-balancing purpose. Organization or ISP with a number of HAs distributedly deployed in many locations then would take considerable benefit from this mechanism. For this Home Agent Relocation mechanism, some simple modification in Registration-Request is needed and a new extension in Registration-Reply is defined.

Modification on Registration-Request. New identification in Registration-Request message is needed to inform HA that the MN performing the registration is a client-access MN. Basically, two ways that can equally be used to inform this presence of client-access type of MN:

- New "C-bit" (Client-Access Mobile-Node bit) is defined using one reserved flag bit in the second byte of Registration-Request message format [1, 2].
- New extension called Client-Access Mobile-Node Extension is defined. As no data is to be included, it is the presence of this extension that matters.

Home-Agent-Relocation Extension. This extension is defined to possibly be included by HA in the Registration-Reply message. Its fields are defined as follows:

- Type : a unique identifier for this extension,
- Length : the length of payload data (in octets),
- Count : the number of alternative HAs listed,
- Home Agents $(1..n)$: a list of alternative HAs suggested by the current HA,
- Preference Levels $(1..n)$: a list of integers greater than 0 that indicates the priority of listed HAs for MN's next registration.

Upon receipt of Registration-Request with Client-Access Mobile-Node bit or extension, HA may return one of these possible responses in its Reply:

1. HA accepts the Request, however it also includes one or more alternative HAs in its HA-Relocation Extension for MN's later use.
2. HA accepts the Request with no alternative HA to be included. This is exactly the same case as an approved registration in base Mobile-IP. For this case, HA includes HA-Relocation Extension with Count field set to 0.
3. HA denies the Request and requires MN to register with one of alternative HA. To indicate this denied registration, a new unique status code should be assigned for Code field of Registration-Reply.

While Home Agent Relocation method allows MN to learn new HAs, it doesn't allocate a required home address for use with each respective HA. The dynamic home address allocation method as specified in [3, 4] can further be employed for this purpose.

2.2 Dynamic Allocation on Foreign Network

Another mechanism needed in our Session-Based Mobile-IP scheme is one that can dynamically allocate HA and home address on the visited foreign network. While the mechanism in [3, 4] is more specifically intended for dynamic allocation on the home network, the basic idea is extendible for that on foreign network as well. In that method, MN sends a Registration-Request with home address field set to 0 to request that one be assigned on home network. Similarly, it can be agreed for our purpose, that a Registration-Request with both home address and HA set to 0 is used to request home address and HA be assigned on the foreign network. As now the authentication can't be done based on home address, Network Access Identifier (NAI) [4] should always be used.

2.3 Processing at MN and FA

In addition to the operations outlined in Mobile-IP specification [1, 2], MN now must take into consideration the fact of its multiple home addresses and HAs. A list of home addresses currently in use needs to be maintained, in which each entry contains information of: its respective HA, binding lifetime, security association, and open-connection list. The open-connection list keeps track of one or more open connections that are currently being bound to a home address. When a new connection is initiated, the current home address must be used for this connection. An entry is then inserted into the open-connection list of this current home address. Conversely, when an application closes a connection, its corresponding entry in the open-connection list must be removed. Should it be the last entry for the corresponding home address, the entry for that home address is removed from the home-address list as well.

Required to support the operations above is a method for determining when an application program initiates and closes a connection. In fact, some techniques for this have been around and employed in various problem domains, such as the one in Network Address Translation [6] for its address binding and unbinding.

Similar to MN, now FA must also take into consideration MN's multiple home addresses. Its visitor list should now be based on MN's NAI instead of home address. For each MN, a list of associated home addresses must be maintained. FA also must differentiate between "primary" and "secondary" MN's registration (a tag can be defined in Request for this). Primary registration must first be successfully performed before MN can get permission for service on the foreign network. Typically this is done with a designated AAA server in MN's home domain and it might involve other AAA-related considerations, such as billing (see Section 5). Following that, some secondary registrations may be performed, with the sole purpose of updating MN's mobility bindings at its HAs.

3   Pre-generated Registration Messages for Fast Handoff

Since the time spans of most sessions are expected to be short to moderate, only a small number of connections would survive for not being closed within the same HA as they are opened. As MN's new point of attachment should remain geographically close to its HAs, there should be no serious problem of long packet delay during re-registration.

Nevertheless, MN that just moved to a new foreign network needs to update its mobility bindings with several HAs. This requires MN to generate a number of needed registration messages during a handoff, which is not so desirable for performance reason, particularly if complex cryptographic operations are in use. To facilitate a fast handoff, here we also describes a technique based on pre-generated registration messages that might be useful.

The basic idea of this technique is that MN prepares all the necessary Registration-Request messages in advance. For this reason, this technique only applies if nonce-based replay protection is in use. Its detailed operation is as follows:

- Whenever possible during idle time, MN pre-generates next Registration-Request messages for all active HAs, except its present HA. The care-of-address field in those messages must be set to MN's present home address.
- Should later MN move to a new foreign network, after it successfully performs registration and gets FA service, MN then registers its newly acquired care-of-address to its last HA. Following that, all the pre-generated Registration messages are sent to other previous HAs.
- When later MN receives IP datagrams from connections with HAs other than last or present HA, MN needs to perform de-tunnelling once (if it's currently using FA's COA) or twice (if it's using a co-located COA) in order to get the original IP datagram.

Figure 1 shows the flow of data to MN, if this technique is employed. One thing to note is that "the last home address" mentioned above needs to be retained in home-address list, as long as there is any connection still using its corresponding HA as second-hop HA.

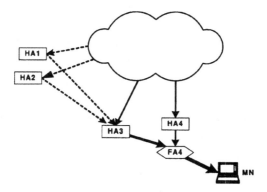

Fig. 1. Data flow to MN using pre-computed Registration messages

4 Some Considerations on Performance

While the new scheme enables MN to obtain new HA on the foreign network, however, it doesn't mean that MN should do so whenever such a HA becomes available. Using too many different HAs can cost MN with undesirable overhead that finally may outweigh the benefit gained.

The question is then how to define a good decision-making procedure that helps MN to determine whether it's still beneficial to obtain a new HA. As this is not an easy task which depends on many factors involved, a configurable setting may be a good strategy. Together with user-supplied values, some heuristics may then be utilized in deciding whether a new HA should be obtained, for example:

- If the number of home addresses hasn't reached a maximum allowed.
- If the number of pending sessions is still below a defined maximum value.
- Periodically, after some time has elapsed since the last HA was obtained.

5 Scalable Authentication Framework

In this section, we outline an authentication framework for securing the operation of our enhancement scheme. Designed to provide a scalable authentication that supports roaming over different organizations and service providers, the authentication protocol is based on public-key cryptography, however its use is kept to a minimum. The protocol outlined here is a further refinement of our earlier work of minimal public-key based authentication for base Mobile-IP [7]. We extend it here by considering the involvement of Authentication, Authorization, and Accounting (AAA) servers, either on foreign domain (called AAAF) or home domain (called AAAH). While our protocol now takes AAA entities in its message flow, it still can equally work should only Mobile-IP standard entities be involved. Furthermore, it introduces no fundamental changes on Mobile-IP operation or its Registration message format.

5.1   Scenario for Authentication

In this scenario, it's assumed that MN and AAAH share a security association. Here, we consider the commonly found case of security association based on shared key, due to its practicality and the fact that MN and AAAH are normally under the same organization. However, we assume that no security association exists between MN and AAAF.

A draft on roaming support of Mobile-IP [8] also considers a similar scenario. While it also suggests the use of public-key based authentication and security association establishment between AAAH and AAAF, the operation is just to rely on mechanisms provided by other infrastructure, namely IPSEC for authentication and IKE for key exchange. However, there are some concerns over that suggestion. Besides its dependency on IPSEC/IKE infrastructure, extra message exchange sessions for IKE in addition to Mobile-IP control messages may result in poor MN's handoffs.

In contrast, our protocol preserves full compatibility of message round with standard Mobile-IP protocol. It makes use of standard extension definition and fully integrates the authentication information with Mobile-IP control messages. This protocol is thus intended to position itself as a more practical yet secure alternative solution, by taking advantage of Mobile-IP properties and trust relationships among its entities. For a more detailed discussion on its design principles and strong points, the reader may also refer to [7].

5.2   The Authentication Protocol

Some notations used:

- MN_{HM} MN's home address;
- MN_{COA} MN's care-of-address;
- HA_{id}, AAA_{id} IP address of HA and AAA server, respectively;
- AAA_{NAI} NAI field of AAA server;
- AAA-$Type$ a value indicating the type of AAA server;
- N_{MN}, N_{HA}, N_{AAAF} nonces issued by MN, HA, and AAAF, respectively;
- $S_{MN\text{--}AAAH}$ shared key between MN and AAAH;
- $\{M\}_K$ encryption of message M under key K;
- $<M>_K$ MAC value of message M (except this MAC field) under key K;
- $Code$ a code value indicating the result of a Request;
- K_{AAAH}, K_{AAAF} public key of AAAH and AAAF, respectively;
- $K^{-1}{}_{AAAH}$, $K^{-1}{}_{AAAF}$ private key of AAAH and AAAF, respectively;
- $<<M>>_{K^{-1}{}_A}$ digital signature of message M (except this field) generated using private key of A;
- $Cert_{AAAH}$, $Cert_{AAAF}$ certificate of AAAH and AAAF, respectively.

The presentation of the protocol here focuses on the message flow and relevant information that must be included in each message. Only message parts following the UDP header are shown. As for the interactions with AAA servers, we just assume the existence of generic AAA entities. Hence, any AAA mechanisms such as RADIUS [10] or DIAMETER [11] can be employed. It's also assumed that the communication between AAA servers and its FA or HA are already secured by the AAA protocol, with its protection of authentication, integrity, and anti-replay. In the protocol run, these secure links are denoted by "$\Rightarrow$".

Our authentication protocol proceeds as follows *(some important fields in each message part are shown)*:

• *Agent Advertisement*

(A1)   FA→MN   : [Mobility Agent Advertisement: MN_{COA}],
[AAAF Identity: $AAAF_{id}$, $AAAF_{NAI}$, AAA-$Type$].

• *Registration Request*

(R1)   MN→FA   : [Registration Request:
$MN_{HM} = 0$, $HA_{id} = 0$, MN_{COA}, N_{HA}, N_{MN}],
[Mobile-Node NAI],
[AAAH Identity: $AAAH_{id}$, $AAAH_{NAI}$, AAA-$Type$],
[Encapsulated FA-Advertisement],
[MN-AAAH Authentication: $<R1>_{S_{MN-AAAH}}$].

where:
[Encapsulated FA-Advertisement] is Message-A1 that's encapsulated in an extension of the T-L-V format. However, two additional fields (8 octets) are also included to carry information of the source and destination IP addresses of the previous Agent Advertisement message.

(R2)   FA   : (upon receipt of R1)
– FA performs all the necessary steps in providing connectivity, such as its visitor list processing for this pending registration [1, 2].
– No authentication steps are required to be done by FA. However, it needs to ensure that the Encapsulated FA-Advertisement is valid as recently advertised. Should FA find it invalid or outdated, FA must cease further processing.

(R3)   FA⇒AAAF   : [Header of AAA protocol for Request command],
[Message in R1].

(R4)   AAAF→AAAH : [Message in R1], $[N_{AAAF}]^2$,
[Allocated Home Address and HA: MN_{HM}, HA_{id}],
[Certificate Extension [9]: $Cert_{AAAF}$],
[AAAF PK-Authentication [9]: $<<R4>>_{K^{-1}_{AAAF}}$].

(R5)   AAAH   : (upon receipt of R4)
– validates $<R1>_{S_{MN-AAAH}}$ using $S_{MN-AAAH}$,
– validates $Cert_{AAAF}$ based on existing PKI in AAAH,
– validates $<<R4>>_{K^{-1}_{AAAF}}$ using authenticated K_{AAAF}.

---

[2] Nonce from AAAF is included, as necessary to prevent replay attack outlined in [7].

- **Registration Reply**

(R6) AAAH→AAAF : [Message R6a], $[N_{AAAF}]$,
[AAAF Identity: $AAAF_{id}$, $AAAF_{NAI}$, $AAA\text{-}Type$],
[Certificate Extension: $Cert_{AAAH}$],
[AAAH PK-Authentication: $<<\text{R6}>>_{K^{-1}_{AAAH}}$].

where:

[Message R6a] : [Registration Reply:
$Code$, MN_{HM}, HA_{id}, N'_{HA}, N_{MN}],
[Mobile-Node NAI],
[MN-AAAH Authentication: $<\text{R6a}>_{S_{MN-AAAH}}$].

(R7) AAAF   : (upon receipt of R6)
- validates N_{AAAF},
- validates $Cert_{AAAH}$ based on existing PKI in AAAF,
- validates $<<\text{R6}>>_{K^{-1}_{AAAH}}$ using authenticated K_{AAAH},
- logs this signed message as a proof of serving MN (perhaps used in conjunction with its billing protocol).

(R8) AAAF⇒FA   : [Header of AAA protocol for Reply command],
[Message R6a].

(R9) FA→MN   : [Message R6a].

(R10) MN   : (upon receipt of R9)
- validates $<\text{R6a}>_{S_{MN-AAAH}}$ using $S_{MN-AAAH}$.

The $AAA\text{-}Type$ in the protocol above indicates the type of AAAF or AAAH. A set of values must be defined for commonly used AAA mechanisms. This field also determines the transport-layer's port number to where the registration messages must be destined in that AAA server.

In the absence of AAAF in the foreign domain, FA can put its own identity in the $AAAF\text{-}Identity$ message extension. Its $AAA\text{-}Type$ field is then set to the reserved value for indicating a Mobile-IP mobility agent. In this case, FA must assume the role of AAAF in performing the necessary authentication steps. Similarly, a normal HA may also function as AAAH. Knowing from the $AAA\text{-}Type$ that the AAAH is actually a HA, AAAF must send the registration message in the standard Mobile-IP message format to the normal Mobile-IP UDP port.

Finally, besides performing authentication, the protocol may additionally distribute new derivative shared-key based security associations between MN and HA, MN and AAAF, or MN and FA. Typically, AAAH generates the keying material and then securely sends it to other parties along with the Registration-Reply message [7, 12]. Afterwards, the relatively faster shared-key based authentication [1, 2] can be used in securing MN's subsequent re-registrations.

6   Conclusion

A new enhancement scheme of Mobile-IP for client-access MN has been proposed in this paper. Some novel ideas are incorporated to give MN better routing and seamless roaming. An authentication protocol is proposed as well, intended to guarantee a secure operation of the scheme. The protocol combines the benefit of scalability provided by public-key cryptography with practicality by keeping computing requirement to a minimum particularly on the MN.

As it's predicted that the demand for Internet mobile access will increase dramatically in the very near future, we hope that our scheme can help facilitate a faster secure Mobile-IP wide deployment.

Current Status

Specification draft is currently underway for contribution to IETF standarization effort, based on the design principles and rationale described in this paper.

Acknowledgements

We are grateful to Mr. Foo Chun Choong (Mobile Computing Group, Dept. of Electrical Engineering, National University of Singapore) for helpful discussions. We also wish to thank the anonymous referees for their constructive comments on the early draft of this paper.

References

1. Perkins, C. E. (ed.): IP Mobility Support. IETF RFC 2002 (1996)
2. Perkins, C. E. (ed.): IP Mobility Support version 2. Internet Draft <draft-ietf-mobileip-v2-00.txt>, work in progress (1997)
3. Calhoun, P. R., Perkins C. E.: Mobile IP Dynamic Home Address Allocation Extensions. Internet Draft <draft-ietf-mobileip-home-addr-alloc-00.txt>, work in progress (1998)
4. Calhoun, P. R., Perkins C. E.: Mobile IP Network Access Identifier Extension. Internet Draft <draft-ietf-mobileip-mn-nai-02.txt>, work in progress (1999)
5. Montenegro, G. (ed.): Reverse Tunneling for Mobile IP. IETF RFC 2344 (1998)
6. Srisuresh, P., Egevang K.: The IP Network Address Translator (NAT). Internet Draft <draft-rfced-info-srisuresh-05.txt>, work in progress (1998)
7. Sufatrio, Lam, K. Y.: Mobile-IP Registration Protocol: A Security Attack and New Secure Minimal Public Key Based Authentication. In: Zomaya, A. Y. et al. (eds.): Proceedings of 4th International Symposium on Parallel Architectures, Algorithms, and Networks (I-SPAN'99), Perth/Fremantle, Australia. IEEE Computer Society, California (1999) 364–369
8. Aboba, B.: Roaming Support in Mobile IP. Internet Draft <draft-ietf-roamops-mobileip-02.txt>, work in progress (1999)
9. Jacobs, S.: Mobile IP Public Key Based Authentication. Internet Draft <draft-jacobs-mobileip-pki-auth-02.txt>, work in progress (1999)
10. Rigney, C., Rubens, A., Simpson, W., Willens, S.: Remote Auth-entication Dial In User Service (RADIUS). IETF RFC 2138 (1997)
11. Calhoun, P. R., Rubens, A. C.: DIAMETER Base Protocol. Internet Draft <draft-calhoun-diameter-08.txt>, work in progress (1999)
12. Perkins, C. E., Calhoun, P. R.: AAA Registration Keys for Mobile IP. Internet Draft <draft-ietf-mobileip-aaa-key-00.txt>, work in progress (1999)

Proxy Agent Consistency Service Based on CORBA

Kyung-Ah Chang, Byung-Rae Lee, Tai-Yun Kim

Dept. of Computer Science & Engineering, Korea University, 1, 5-ga, Anam-dong,
Sungbuk-ku, Seoul, 136-701, Korea
gypsy93, brlee, tykim@netlab.korea.ac.kr

Abstract. Mobile agent [1] platforms use proxy agents to provide location transparent communication. However, these platforms lack of consistency service between proxy agents and mobile agents. In this paper, we propose CORBA [2]-based proxy agent consistency service with MASIF [1] naming services. Control messages are defined to warn, request, update and confirm consistency related activities. If naming service is not available, proposed service can still update the proxy agent with up-to-date IOR [2] information. For reliable agent messaging, message forwarding is possible through the proposed proxy agent consistency service.

1 Introduction

One of the most compelling visions of the future is a world in which specialized agents collaborates with each other to achieve a goal that an individual agent cannot achieve on its own. These agents would need to move, when appropriate, to conduct high-bandwidth conversations that could not possibly take place over a low-bandwidth network. A platform [3–8] for building these systems, therefore, would have to support a location transparent communication system that allows mobile agents to communicate seamlessly with each other, even if they are moving across a network.

Mobile agent platforms (e.g. mobile agents, Aglet [5,6] and Voyager [7,8] based on Java [10]) use proxy agent concept to provide location transparent interaction with each other. But, these platforms lack of the consistency support between proxy agents and its real mobile agents. For location transparent interaction, efficient and fault tolerant proxy agent consistency service should be provided.

The recent OMG work on a Mobile Agent System Interoperability Facility (MASIF) [1] specification can be regarded as a milestone on the road toward a unified distributed mobile object middleware, which enables technology and location transparent interactions between static and mobile objects. In order to support both the traditional client/server paradigm and mobile agent technology, take advantage of interoperability with further agent platforms of different manufacturer, compliance to the OMG MASIF standards has to be achieved.

In this paper, we propose CORBA-based proxy agent consistency service with control messages which provides a means for consistency between proxy

H.V. Leong et al. (Eds.), MDA'99, LNCS 1748, pp. 230–239, 1999.
© Springer-Verlag Berlin Heidelberg 1999

agent and real mobile agent. To provide interoperability with other mobile agent platforms, MASIF compliance is achieved. In case of MAFFinder [1] or home agent system [1] failure, proposed service provides a means to update proxy agent with up-to-date Interoperable Object Reference (IOR) [2] information of the target mobile agent. Messages arrived at the target agent's old location can be forwarded to the new location through the proposed service.

The remainder of this paper will proceed as follows. In Sect. 2, we describe location transparency mechanism in various mobile agent systems and MASIF naming service. In Sect. 3, CORBA-based proxy agent consistency service is presented. Control messages used in consistency service is explained and detail proxy agent consistency service in MASIF are presented. In Sect. 4, we show the performance analysis. Finally, in Sect. 5, we give the conclusion and future work to this work.

2   Location Transparency

In this section, we describe location transparency in other mobile agent systems and MASIF naming service. We also study about the benefits of integrating mobile agent technology and CORBA. MASIF solutions of CORBA Naming Service [10] problems will be explained subsequently.

2.1   Proxy Agent

A server agent is an object that can exist outside the local address space of an application. An application can communicate with a server agent by constructing a virtual version of the server agent locally. This virtual version is called a proxy agent and acts as a reference to the server agent. When messages are sent to a proxy agent, the proxy agent forwards the messages to the server agent. If an agent moves from one application to another, you can still locate the agent by using its last known address [5–8].

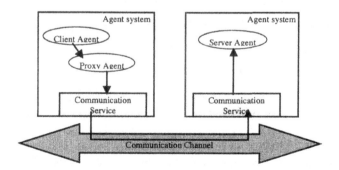

Fig. 1. Location transparent communication using proxy agent

Figure 1 shows proxy agent based location transparent service in mobile agent platforms. Aglets and other mobile agent platforms use similar approach.

2.2   MASIF

The CORBA has been established as an important standard, enhancing the original Remote Procedure Call (RPC) based architectures by allowing relatively free and transparent distribution of service functionality [11, 12]. Besides mobile agent technology has been proved to be suitable for the improvement of today's distributed systems. Due to its benefits, such as dynamic, on-demand provision and distribution of services, reduction of network traffic and the reduction of dependence regarding server failures, various problems and inefficiencies of today's client/server architectures can be handled by means of this new paradigm. However, for several applications RPCs still represent a powerful and efficient solution. Thus, an integrated approach is desirable, combining the benefits of both client/server and mobile agent technology [13].

Mobile agent technology is driven by a variety of different approaches regarding implementation languages, protocols, platform architectures and functionality. MASIF is developed to achieve a certain degree of interoperability between mobile agent platforms of different manufactures.

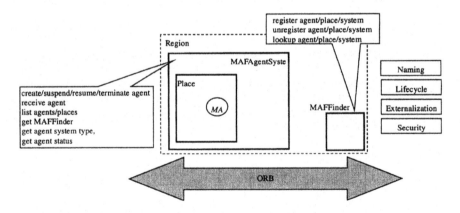

Fig. 2. The MASIF architecture

As shown in Fig. 2, MASIF has adopted the concepts of places and agent systems that are used by various existing agent platforms. Two interfaces are specified by the MASIF standard: the MAFAgentSystem [1] interface provides operations for the management and transfer of agents, whereas the MAFFinder [1] interface supports the localization of agents, agent systems, and places in the scope of a region or the whole environment, respectively.

2.3   Location Transparency in MASIF

The CORBA Services are designed for static objects. When CORBA naming services, for example, are applied to mobile agents, they may not handle all cases as well. The MAFFinder functions as an interface of a dynamic name and location database of agents, places, and agent systems.

Stationary agents as well as mobile agents may publish themselves, since a CORBA object reference (IOR) [2] comprises, among others, the name of the host on which an object resides and the corresponding port number, a mobile agent gets a new IOR after each migration. In this case, the IOR that is kept by the accessing application becomes invalid. Following three solutions for this problem are specified in MASIF.

1. The first solution is that the ORB itself is responsible for keeping the IOR of moving objects constant. The mapping of the original IOR to the actual IOR of the migrated agent is managed by a corresponding proxy object, which is maintained by the ORB. Although this capability is described by CORBA, it is not a mandatory feature of an ORB. Thus, the MASIF standard does not rely on this feature [1].
2. The second solution is to update the name IOR associated to the mobile agent after each migration, i.e. to supply the Naming Service with the actual agent IOR. This can be done by the agent systems, which are involved in the migration process or by the migrating agent itself. In this way, the naming service maintains the actual IOR during the whole lifetime of the agent. If an application tries to access the agent after the agent has changed its location, the application retrieves an exception (e.g. invalid object reference). In this case the application contacts the Naming Service in order to get the new agent IOR. A disadvantage of this solution is that the MAFFinder must be available every time in order to retrieve the new IOR to which each message must be sent [1].
3. When a mobile agent migrates for the first time, the original instance remains at the home agent system and forwards each additional access to the migrated instance at the new location. In this way, the original IOR remains valid, and the clients accessing the agent need not care about tracking it. They still interact with the original instance, called proxy agent, which only exists to forward requests to the actual (migrating) agent. One disadvantage is that the home agent system must be accessible at any time. If the home agent system is terminated, the agent cannot be accessed anymore, since the actual IOR is only maintained by the proxy agent [1].

To provide overcome weaknesses of MASIF and provide efficient and reliable location transparent interaction, following main objectives of the development should be achieved:

- In order to take advantage of interoperability with further agent platforms of different manufacturer, compliance to the OMG MASIF standards has to be achieved.

- For reliability, messages should be guaranteed to arrive at the designated agent system.
- Even in the home agent system or MAFFinder failure, proxy agent consistency service should be provided.

3   Proxy Agent Consistency Service Based on CORBA

In this section, CORBA-based proxy agent consistency service is presented. Proposed service can be applied over MASIF solutions, resulting in efficient MASIF compliant proxy agent service. Proposed service supports the second and third solution of MASIF, MAFFinder and home agent system based concepts, because MASIF doesn't rely on the first solution, keeping IOR to the actual agent IOR by ORB.

3.1   Control Messages

In the proposed proxy agent consistency service, control messages are defined to provide a flexible mechanism for agent systems to update their proxy agents. It is the mechanism by which mobile agent request forwarding services after migration, keeping agent systems' proxy agents be up-to-date information.

Agent Warning message

It is a CORBA exception (invalid object reference).

Agent Request message

- It is a simply a *lookup_agent()* operation of MAFFinder.
- It is a request of an IOR of the target mobile agent.

Agent Update message

It is a response to *Agent Request* message. It contains the IOR of the target mobile agent.

Agent Confirm message

It is used to acknowledge receipt of an *Agent Update* message.

3.2   System Architecture

To process control messages, some modules are added to the original agent system. The agent system contains MASIF Core Service module for interoperability with other mobile agent platforms. In the absence of control message facility, a agent system can contact MAFFinder to find out a current IOR of the target mobile agent, or messages destined for a mobile agent will be routed to the home agent system in the same way as defined in MASIF. Message Management, Proxy Agent Management modules are for the control messages processing. Message Queue module guarantees the reliable message delivery mechanism through message forwarding.

- MASIF Core Service: For interoperation with other mobile agent platforms, it implements MASIF standard interfaces.
- Message Management: It accepts and transfers control messages to update proxy agents. After receiving control messages, it delivers the message to Proxy Agent Management module to update proxy agent connection.
- Proxy Agent Management: It receives control information from the Message Management module, and gives order to update the proxy agent. It creates, modify and destroy proxy agents.

3.3 Proxy Agent Consistency Service Based on MASIF Solution

When the naming service policy is based on the MASIF's second solution, MAFFinder based approach, following methods can be applied over the MASIF's solution to update proxy agents.

When any agent system receives a transferred message, if it has a proxy agent for the target mobile agent, this agent system may deduce that the source agent system has an out-of-date proxy agent entry for the target mobile agent. In this case, the receiving agent system should send an Agent Warning Message to the source agent system, advising it to contact the MAFFinder to find out target mobile agent's current IOR.

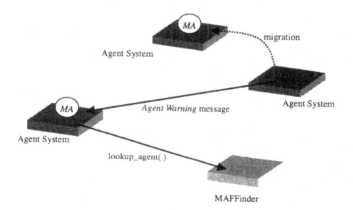

Fig. 3. Proxy agent consistency service in MASIF naming service 2

When the naming service is based on the MASIF's third solution, home agent system based approach, following method can be applied over the MASIF's solution to update proxy agents.

When any agent system receives a transferred message, if it has a proxy agent for the target mobile agent, this agent system may deduce that the source agent system has an out-of-date proxy agent for the target mobile agent. In this case, as shown in Fig. 4, the receiving agent system should send an *Agent Warning* message to the agent system, advising it to send an *Agent Request* message to the home agent system of the target mobile agent. The source agent system can transfer *Agent Request* message to the home agent system when it needs to update the proxy agent. After receiving the *Agent Request* message, the home agent system can transfer the *Agent Update* message to the source agent system.

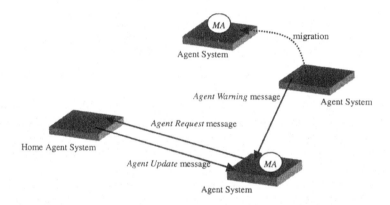

Fig. 4. Proxy agent consistency service in MASIF solution 3

3.4   Proxy Agent Consistency Service for Message Forwarding

In case of agent system or MAFFinder failure, proposed system can still provide proxy agent consistency service.

Messages in flight transferred to the old location when the mobile agent moved are likely to be lost and are assumed to be retransferred if needed. To avoid risks of message losses, proposed service provides a means for the mobile agent's previous agent system to be reliably notified of the mobile agent's new IOR, allowing messages in flight to the mobile agent's previous agent system to be forwarded to its new location.

As shown in Fig. 5, the mobile agent need to transfer *Agent Update* message to its previous foreign agent system until the matching *Agent Confirm* message is received. Previous agent system can forward messages to the exact agent system using the IOR information from the *Agent Update* message. This notification is for any messages transferred to the target mobile agent's previous agent system

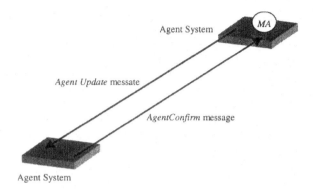

Fig. 5. Proxy Agent Consistency Service for Message Forwarding

from correspondent mobile agents with out-of-date proxy agent, to be forwarded
to its new agent system.

4   Performance Analysis

In this section, we provide performance comparison between proposed proxy
agent consistency service and other mobile agent platforms. Subsequently, performance comparison with MASIF is described.

4.1   Comparison with Mobile Agent Platforms

Integration of CORBA and mobile agent technology is desirable, combining the
benefits of both client/server and mobile agent technology, and on the other
hand eliminating or at least minimizing the problems that rise if one of these
techniques is used as "stand-alone" solution.

Table 1 shows the performance analysis of other mobile agent platforms and
proposed CORBA based remote messaging mechanism.

Table 1. Performance Analysis with Mobile Agent Platforms

	Ara [4]	Aglet [5]	Voyager [8]	Proposed Mechanism
Consistency Service for Message Forwarding	Not stated	Not stated	O	O
Interoperability (MASIF-compliant)	×	×	×	O
Language Independence	△	×	×	O
Proxy Agent Tracking	△	O	O	O

×: not supported, △: weakly supported, O: well supported

Only proposed mechanism is MASIF compliant. Although Voyager is CORBA enabled, it does not conform to MASIF.

Ara is dependent on TCL, Aglet and Voyager are implemented in Java. Although Voyager is based on CORBA, it is Java dependent. Proposed proxy agent consistency service conforms to MASIF, so language independence is acquired.

Both Ara and Aglet include simple, local lookup mechanism that enables associating a string with an agent URL. Voyager contains its own directory service. These platforms do not support MASIF naming service. Proposed CORBA based remote messaging mechanism use CORBA Naming Service and MAFFinder to provide naming service in large distributed agent environment.

4.2   Analysis with MASIF

Table 2 shows performance enhancements of proxy agent consistency service comparing with MASIF. MASIF does not provide any proxy agent service. Proposed proxy agent service provides location transparent operation among mobile agents. It provides proxy agent tracking service with control messages. In case of MAFFinder or home agent system failure, proposed system can still update proxy agents with up-to-date IOR information through *Agent Update* message.

Table 2. Performance analysis with MASIF

	MASIF solution 2	MASIF solution 3	Proposed scheme
Consistency Service for Message Forwarding	Exception (Retransmission)	No specification	Forwarding
Consistency Service in MAFFinder Failure	Not Supported	Supported	Supported
Consistency Service in Home Agent System Failure	Supported	Not Supported	Supported
Proxy Agent Tracking	Not Supported	Not Supported	Supported

To avoid risks of message losses, proposed CORBA-based proxy consistency service provides a means for the mobile agent's previous agent system to be reliably notified of the mobile agent's new IOR, allowing messages in flight to the mobile agent's previous agent system to be forwarded to its new location.

5   Conclusion and Future Work

We proposed CORBA-based proxy agent consistency service with control messages, which provides a means for consistency between proxy agent and real mobile agent. To provide interoperability with other mobile agent platforms, proposed service could be used in MASIF environment. Fault tolerant, reliable proxy agent services can be achieved through the proposed service.

As future work, we are considering secure control messages. Control messages should not be forged, destroyed illegally. Integrity and confidentiality of control messages should be provided for secure agent collaboration.

References

1. OMG, Mobile Agent System Interoperability Facilities Specification, November, 1997.
2. OMG, The Common Object Request Broker: Architecture and Specification, 1998.
3. General Magic, Odyssey Web Site, URL: http://www.genmagic.com/agents/, 1999.
4. Concordia, Mitsubishi Electric Information Technology Center America, Horizon Systems Laboratory, http://www.meitca.com/HSL/Projects/Concordia/-Welcome.html, 1999.
5. IBM Tokyo Research Labs. Aglets Workbench: Programming Mobile Agents in Java, http://www.trl.ibm.co.jp/aglets/, 1999.
6. Danny B. Lange and Mitsuru Oshima, Programming and Deploying JavaTM Mobile Agents with AgletsTM, Addison Wesley, 1998.
7. ObjectSpace, Voyager Core Package Technical Overview, 1997.
8. ObjectSpace, ObjectSpace Voyager CORBA Integration Technical Overview, 1997.
9. Java Team, James Gosling, Bill Joy, and Guy Steele, The JavaTM Language Specification, Sun Microsystems, 1996.
10. OMG, The Common Object Request Broker Service, 1998.
11. Andreas Vogel and Keith Duddy, JAVATM Programming with CORBA, Wiley, 1998.
12. Robert Orfail, Dan Harkey and Jeri Edward, Client/Server Survival/Guide, 3rd Ed., Wiley, 1999.
13. T. Magedanz, K. Rothermel, and S. Krause, "Intelligent Agents: An Emerging Technology for Next Generation Telecommunications?," Proceedings of IEEE IN-FOCOM '96, pp. 464–472, IEEE Press, 1996.

Keynote Speeches

Integrating Mobile Objects into the Wirelessly Wired World: The Need for Energy Efficient Algorithms

Prof. Imrich Chlamtac

University of Texas, Dallas

Abstract. There is a need to design a new type of communication systems which can respond to the restricted and changing energy resources encountered due to mobility. Energy efficient or "Power Aware" Communication needs to create protocol building blocks that meet given Quality of Service and performance criteria, while working within a limited energy budget. Devices and systems such as smart tags, smart pagers, intelligent ID cards, warehouse identification tags, store labels, personal battlefield communicators, have dealt with the issue of power for several years now. More recently consumer electronics has entered this realm too. New Internet ready "smart phones", expected to appear in the market by mid 2001, will adapt to changes in frequencies, transmission standards and protocols around the world at a push of a button. These "software defined radios" will provide a universal way for connecting to multimedia services and Internet. In all these, the battery power to support them is seen as a primary limitation.

In this talk we present some of the new paradigms leading to the design of communications protocols with energy constraint. Their general principle is the minimization of time in which the device needs to be active (awake), the maximization of the limited bandwidth utilization, and meeting access delays of the applications' constraints. We will use tag type devices as a vehicle to present solutions available at the access/network layer. Through these we will demonstrate approaches unique to the energy aware communication as it moves from tags, to phones, to desktop to portable and mobile platforms by efficiently dealing with power-aware needs and strategies.

Databases Unplugged: Challenges in Ubiquitous Data Management

Prof. Michael Franklin

University of California, Berkeley

Abstract. The recent Asilomar Report on Database Research cites ubiquitous "information appliances" as a major driver for database systems research over the next ten years. For the present, however, the use of most mobile devices is mostly restricted to Personal Information Management (PIM) applications. In order to fully realize the potential of ubiquitous data access, such devices must be allowed to serve as an extension to the enterprise data management infrastructure. Thus, database technology must be adapted to deal with the limitations as well as the tremendous opportunities of large-scale mobile data access. Challenges exist in core database areas such as system architecture, query processing, and transaction management, as well as in emerging areas such as data dissemination and user-directed execution. In this talk I will outline several of these challenges and describe approaches that have been proposed to meet them.

Author Index

Lecture Notes in Computer Science

For information about Vols. 1–1676
please contact your bookseller or Springer-Verlag